Ethnobiology and the Science of Humankind

ETHNOBIOLOGY AND THE SCIENCE OF HUMANKIND

Journal of the Royal Anthropological
Institute Special Issue No 1

EDITED BY ROY ELLEN

 Blackwell Publishing

 Royal Anthropological Institute

© Royal Anthropological Institute of Great Britain & Ireland 2006

BLACKWELL PUBLISHING
350 Main Street, Malden, MA 02148-5020, USA
9600 Garsington Road, Oxford OX4 2DQ, UK
550 Swanston Street, Carlton, Victoria 3053, Australia
Royal Anthropological Institute, 50 Fitzroy Street, London W1T 5BT

The right of Roy Ellen to be identified as the Author of the Editorial Material in this Work has been asserted in accordance with the UK Copyright, Designs, and Patents Act 1988.

All rights reserved. No part of this publication may be reproduced, stored in a retrieval system, or transmitted, in any form or by any means, electronic, mechanical, photocopying, recording or otherwise, except as permitted by the UK Copyright, Designs, and Patents Act 1988, without the prior permission of the publisher.

First published 2006 by Blackwell Publishing Ltd

Library of Congress Cataloging-in-Publication Data has been applied for

ISBN 10: 1-4051-4589-7
ISBN 13: 978-1-4051-4589-7

A catalogue record for this title is available from the British Library.

Set in Hong Kong
by SNP Best-set Typesetter Ltd.

The publisher's policy is to use permanent paper from mills that operate a sustainable forestry policy, and which has been manufactured from pulp processed using acid-free and elementary chlorine-free practices. Furthermore, the publisher ensures that the text paper and cover board used have met acceptable environmental accreditation standards.

For further information on
Blackwell Publishing, visit our website:
www.blackwellpublishing.com

Contents

Preface		vi
1	Introduction Roy Ellen	1
2	The First Congress of Ethnozoological Nomenclature Brent Berlin	29
3	Ethnobiology and the evolution of the human mind Steven Mithen	55
4	The interplay of ethnographic and archaeological knowledge in the study of past human subsistence in the tropics David R. Harris	77
5	Amazonian historical ecologies Laura Rival	97
6	The interface between medical anthropology and medical ethnobiology Anna Waldstein & Cameron Adams	117
7	Ethnobiology and applied anthropology: rapprochement of the academic with the practical Paul Sillitoe	147
8	Meeting of minds: how do we share our appreciation of traditional environmental knowledge? Eugene Hunn	177
Index		197

Preface

Ethnobiology and the science of humankind is a particularly appropriate volume with which to launch the new *JRAI* special issue series. It seeks to address problems of the widest concern in anthropology as represented in the mission of the Royal Anthropological Institute – from biology to reflexivity – and it attempts to achieve a genuine integration of purpose. The collection as a whole promises to provide a benchmark assessment of the subject of ethnobiology in relation to the broader concerns of anthropologists of various sub-disciplinary affiliations.

The volume is derived from a flagship panel hosted by the Ninth International Congress of Ethnobiology held at the University of Kent at Canterbury in June 2004. This meeting was held jointly with the 45th Annual Meeting of the Society for Economic Botany and the Eighth International Congress of Ethnopharmacology. Given the title of the panel it is appropriate that the Congress should have been sponsored by the RAI (which was able to fund a number of student bursaries), and by the Wenner-Gren Foundation, which has always had a refreshingly broad and un-sectarian definition of anthropology as the multidisciplinary sciences of humankind. The panel itself received British Academy support, for which we are most grateful. I am also grateful to a number of individuals for making this volume happen, either as facilitators or with editorial assistance: the members of the RAI Publications Committee and particularly Jeremy MacClancy, who has acted as a liaison, Hilary Callan, Simon Platten, Paul Sillitoe, Sarah Johns, Brian Morris, Susi Soemarwoto, and Glenn Bowman. Finally, the editing would not have been completed on time without the financial assistance of the Centre for Computing and Social Anthropology at Kent and ESRC grant RES-000-22-1106.

Roy Ellen

1

Introduction

ROY ELLEN

Background
This volume explores the contribution of recent work in ethnobiology[1] to anthropological insights in the widest sense. As a project, it arises from the observation that, increasingly, the subject matter and methodologies of ethnobiological research address core questions about the character of culture, language, cognition, knowledge, and human subsistence, and how these interact through, for example, long-term processes of co-evolution. We seek to provide here some kind of qualitative test of the assertion that ethnobiology stands at an important intellectual junction between biology, culture, and sociality; a view which the authors of this collection share, but which is not necessarily or always apparent in the practice of individual exponents. That ethnobiology has not always been seen to occupy such a critical position also needs explaining. In order to meet such a requirement, this introduction supplies some historical background to developments that have taken place over the last fifty years or thereabouts. We are, however, mostly concerned with current work, and indeed with the prospects for future research in this field of study. We offer, therefore, a retrospective certainly, but also and most importantly, a tentative prospective.

Richard Ford (1978) once memorably said that ethnobotany – but we can extend the point to ethnobiology as a whole – represented a common discourse but lacked a unifying theory. Since that observation was made things have changed. Certainly, ethnobotany, at least, is now replete with methods manuals

(Alexiades 1996; Martin 1995), readers (Minnis 2000), and textbooks (Cotton 1996), as well as taught programmes around the world. While methods, textbooks, and courses are not in themselves evidence of theory, they do imply the attainment of a critical level of agreement as to the issues which ethnobiology should address, which must surely be a precondition for the development of theory. And there are some excellent recent overviews, such as those of Miguel Alexiades (2003), Gary Martin (2002), Daniel Clément (1998a; 1998b), Doug Medin and Scott Atran (1999), Stanford Zent (n.d.), and others, which have identified a degree of coherence and a defined trajectory over time which those in the pioneer phase might now find surprising. It is certainly a more visible subject. Whether there is now something which we might describe as distinctive ethnobiological theory is another question, but the subject is certainly more theorized, from a number of perspectives.

The history of ethnobiology as a discourse has been documented elsewhere (Brown 1984: 2; Bulmer 1975; Clément 1998a; Ellen 1986; Hays 1974; Porteres 1977; Sturtevant 1964; Zent n.d.). In histories and general accounts of both ethnobiology and ethnobotany it has become conventional to distinguish two dimensions, orientations or levels (Bulmer 1975: 10), or phases (Hays 1974: 100-3), of inquiry (also Davis 1995: 43; Ellen 1996: 457-8). The first phase was typified by the determination of culturally, or more specifically economically, significant species and their assessment in terms of 'the objective biological dimension'. At its earliest and most rudimentary, this comprised the listing of names and uses of plants and animals in native non-Western or 'traditional' populations, often in the context of salvage ethnography; or, more crudely put, *ethno*-biology as the descriptive biological knowledge of 'primitive' peoples (e.g. Castetter 1944; Harrington 1947; Stevenson 1915). In the second phase or dimension, which emerged historically and logically from the first, the focus was on the study of human conceptualization and classification of the natural world, a development in retrospect, and iconically, marked by the appearance of Harold Conklin's doctorate in 1954. But in both of its senses, and for most of its history, ethnobiology has usually been a secondary adjunct to other studies, and the importance of such a service should not be understated. In this ancillary role the distinction between the two orientations is perhaps understandable, even necessary, but increasingly these have become aspects of a single problematic. Thus, in the British tradition, anthropological ethnobiology (meaning largely, but not exclusively, studies of folk biological classification) acquired what legitimation it has in its liberation by Claude Lévi-Strauss (1966) from being merely a branch of linguistics and folklore (Bulmer 1975: 10), but it has subsequently become much more than this. In France, ethnobiology needed no such legitimation, with independent roots in the linguistic work of André Haudricourt (e.g. 1973; Haudricourt & Hédin 1987

[1943]) and in a vigorous local tradition of economic biology (Clément 1998a; 1998b; Porteres 1977).

Today, ethnobiology is, first and foremost, the study of how people of all, and of any, cultural tradition interpret, conceptualize, represent, cope with, utilize, and generally manage their knowledge of those domains of environmental experience which encompass living organisms, and whose scientific study we demarcate as botany, zoology, and ecology. (For an alternative definition see Clément 1998a: 19.) But ethnobiology – like anthropology more broadly – seeks to go beyond the local, to compare such knowledge and its consequences between different human populations, and to establish generalizations that are valid at the regional, global, and species level. In order to do this it must inevitably translate into the special-purpose categories, language, and intellectual issues of global (no longer purely Occidental) scholarship.

Ethnobiology is now as much analytic as it is descriptive, and has begun to develop conspicuously its own theory. Like ethnography, it has made a virtue of its practice, and like social anthropology, it is as much defined by its methods as by its theory. Areas where ethnobiological methods and theory have become particularly distinctive include: resource pool approaches; quantitative plot studies; the links between biological and cultural diversity (including agrobiodiversity); the historicizing of global biological exchange; resource sustainability; the problematizing of non-timber forest product issues; knowledge transmission and erosion; valuation theory; the comparative study of the bioactivity of useful species in relation to taxonomy and cultural convergence; the concept of ethnobiological keystone species – to say nothing of all that theory which came with the examination of ethnobiological classification, and which Brent Berlin has done so much to advance. Despite all this, ethnobiology has conventionally been seen by its practitioners as only an interstitial subject, and has been so perceived by outsiders: peripheral, interdisciplinary, and derivative, an importer rather than an exporter of ideas and techniques. The present volume seeks to explore the contention that, in fact, there has been an equally significant counter-flow of ideas and practices from ethnobiology, and is concerned with that outflow of contributions into anthropology more generally. It additionally seeks to make the case for ethnobiology as a distinctive interdisciplinary subject which is especially suited to developing synergies which go beyond the 'mixing' of ideas from adjacent and overlapping subjects. There is, of course, a persuasive case to be argued in favour of the influence of ethnobiology on other subjects: ecology, for example (Balée 1998; Plotkin 1995; Schultes & von Reis 1995: part 6), ethnopharmacology and ethnopharmacy (Etkin 1986; 1988; 1993; Schultes & von Reis 1995: part 8), conservation biology (Cunningham 2001; Johannes 1989; Laird 2002; Tuxill & Nabhan 2001), development studies (Warren, Slikkerveer & Brokensha 1995), political ecology

(Zerner 2000), and economic botany,[2] being those that most readily come to mind; and even for its intrinsic proto-disciplinarity (Alexiades 2003); but here we concentrate on the contribution to anthropology, where connections at the interface have perhaps been best articulated.

This introduction attempts to provide, as it were, an empirical account of the kinds of theories which ethnobiologists employ, where they come from, and the extent to which they might constitute shared, and recognizably anthropological, theory. The conjunctural qualities that I attribute to ethnobiological research arise in part because, like many other fields of 'the science of humankind', it seeks to produce knowledge of the relationship of categories to behaviour, and of culture to social action; but it is additionally unusual in having tangible material referents which are simultaneously the units of analysis in natural science discourse. A comparable simultaneity of reference has been argued for studies of material culture, which Sillitoe suggests also provide us with 'a sure observational baseline' (1988: 5); and in both cases the intersection between the discourses provides opportunities to test the rigour of our methods through repeated and systematic cross-reference. The introductory remarks which follow are restricted to just seven headings, are in no sense either exhaustive or mutually exclusive,[3] but do have a certain salience in the literature: the foundational paradigm of taxonomic orthodoxy – what I call 'the Linnean grid'; language and the translation of knowledge systems; cognition and culture; the social organization and transmission of knowledge; medical ethnobiology; the applied practice of ethnobiology; and – the meta-theory which binds all this together – the co-evolutionary paradigm as part of a 'biocultural synthesis'.

The Linnean grid

What links the work of most ethnobiologists, whatever other theoretical orientations they may subscribe to, is at least passing acknowledgement of the relevance of contemporary biological (and specifically taxonomic) orthodoxy. Empirical work which has paid no attention to this can be of no lasting value as its phylogenetic reference will always be in doubt. If descriptions of the natural world are rendered entirely in terms of local words and categories, they will be of virtually no use to other scientists, other local folk, and policy advisers who must work within a world of firm identifications and frameworks. Of course, the distinction between indigenous (or local, or traditional) and modern (or scientific) knowledge is now much discussed and critiqued, and there is no settled agreement on the terms we use to express the differences which such a debate encodes (Ellen & Harris 2000), but science itself and the pragmatics of intercultural communication and action require that provisional working assumptions be made.

For some ethnobiologists the Linnean scheme provides no more than a grid on which to map folk categories, a way of pinning them down to some more widely shared representations; for others that grid is a crucial part of an argument, either by demonstrating the degree to which folk classifications might match or deviate from their scientific counterparts in terms of category boundaries or representations of diversity, or by demonstrating the cultural significance of biological information. But some people who call themselves ethnobiologists also operate wholly within a biological paradigm, being concerned with the uses of a particular species, biological family, or functional group of species, rather than with their local cultural representation and social correlates. Some of these individuals would regard themselves as anthropologists, but others have a primary allegiance to a different discipline. Consult any issue of *Economic Botany, Journal of Ethnopharmacology, Human Ecology*, or *Journal d'Agriculture Tropicale et Botanique Appliquée* for insight into work of this kind. At a more analytical and inclusive level such an approach informs those who seek to place the understanding of human subsistence strategies generally in some overarching biological paradigm: say, diet breadth studies, evolutionary ecology, optimal foraging theory, or pharmacognosy.

If anthropologists wish to pride themselves on their ability to apprehend and translate the ethnographic 'other', then they should perhaps with equal commitment take on board translation between the categories of international science and folk knowledge. There is a parallel case here with all anthropology, of course, in that we write in English – or in some other widely spoken tongue of a nation in which global science is instituted – and so translate between technical and folk categories all the time, even if they are anthropological ones (e.g. clan, shaman, taboo). In his contribution to this volume, Eugene Hunn – while acknowledging the pragmatic necessity of 'Latinate names' – attempts precisely this. He invites us to speculate reflexively on the apparent arbitrariness of the Linnean grid. This, especially in the context of the 'new systematics' based on cladistics, phylocodes, molecular evidence, and fear of species overload, is increasingly generating patterns which fail to match in convenient ways the kinds of resemblances enshrined in a taxonomic orthodoxy based on herbarium sheets, spirit specimens, skins, and osteology. We can see this in botany (Hollingsworth, Bateman & Gornall 1999), zoology (Pennisi 2001), and indeed in primatology and evolutionary anthropology (Relethford 2003: 248-53). Hunn suggests that we might consider scientific Latin as just one 'special-purpose' language amongst many capable of grasping the reality of biodiversity, and in examining the often technically profound way in which many unwritten languages of local peoples encode biological knowledge the evidence is plain to see. That Latin and not Tzeltal has become the global tool for accurately describing organisms is due to the vagaries of history: the adoption of

Latin as the language of the Roman church, and therefore of European scholarship, the expansion of Europe into the rest of the world rather than the other way around, and the eighteenth-century scientific enlightenment (Stearn 1973: 6-50).

Language and the translation of knowledge systems

Early work in ethnobiology often addressed the problems of lexicographers and philologists. This was especially the case in France (e.g. Haudricourt 1973; see also Bulmer 1974: 79). But the approach derived from philology, and indeed from modern historical linguistics, is still reflected in the use of ethnobiological language data to track the introduction, erosion, extinction, diffusion, domestication, and change in significance of useful species (Balée 1994; Nabhan & Rea 1987; Whistler 1991), and to correlate biodiversity with cultural diversity using language as a proxy (Maffi 2001). Moreover, modern ethnobiology (which, if we take the longer view, begins seriously in about 1950 with the emergence of what I earlier called 'the second phase') found its first theoretical stimulus in linguistics, either in the anthropological linguistics of Franz Boas, Edward Sapir, and Benjamin Lee Whorf or (later) in the structuralism of Roman Jakobson and Ferdinand de Saussure. The linguistic methods employed during this formative phase are sometimes labelled ethnosemantic (or at least were in its first wave), and in the United States owed much to work published in the late 1950s and early 1960s in the ethnoscience tradition (Sturtevant 1964), the guiding methodological principle of which was to yield sufficient data to understand those rules which might permit an ethnographer successfully to replicate the language behaviour of a native (e.g. Frake 1980).

The combined methodologies of ethnoscience and componential analysis proved to be a productive paradigm in terms of the studies they inspired, and influenced the analytical terminologies of a later generation. But while they served well as schemata for yielding basic data, and their elicitation techniques are still valuable today, their fatal drawbacks included over-reliance on formal methods of interview, a preoccupation with nomenclature and distinctive feature analysis to the exclusion of much else, accompanied by an over-simplified sociology, and a tendency ultimately to play down the dynamics of sharing knowledge and cross-cutting rules of classifying behaviour in favour of eliciting taxonomic schemes of often bewildering complexity. In addition to its techniques, its lasting value also lay in its concern with the relationship between category and word; and an ability to show that the correspondence between the two was seldom straightforward, for example that not all words imply the existence of separate categories, and that categories could exist independent of lexical labels. This was made possible in part by the mapping of folk categories on to their phylogenetic denotata: thus theorizing the relationship between emic and etic based on reproducible empirical data. Such a linguistic approach is perhaps best exemplified in the work of Harold Conklin (1954) and continues, for example, in the work of Paul Taylor (1990). It is grounded in the assertion that since biological classification is a part of language, then it must necessarily be understood primarily as a linguistic phenomenon, employing the techniques of linguists.

Hunn emphasizes here the importance of understanding local languages for a full appreciation of local ethnobiology. Names of fauna and flora have also been crucial in the study of onomatopoeia, metaphor, and sound symbolism, which both he and Berlin refer to in their respective papers. But Hunn also reminds us that language, and especially written language, has its limits. Not only is the written word often inadequate to grasp the precise way in which local peoples perceive their environment, but the vocal and verbal dimension itself is insufficient. Despite the attempts of anthropologists such as Steven Feld (1996) and Paul Stoller (1989), the synaesthetic reality of sensory perception of the environment can be reduced to written texts only with difficulty, and this is partly the reason why it is so hard to reduce practical or embodied ethnobiological knowledge to a written text. Hunn suggests that some of this reality is captured in the often dazzling illustrations which accompany ethnobiological texts, but he also points to a future in which multimedia ethnographies (see, e.g., Hesse-Biber, Dupuis & Scott Kinder 1997) will allow us to extend the realms of what is possible to publish as scientific and scholarly communications. Moreover, recent cognitive anthropology, as reflected below, and in the chapter here by Steven Mithen, has now effectively demonstrated the ability of the mind to make sense of much ecological knowledge, and indeed culturally to transmit such data, without constantly converting it into language (Ellen 2003*a*: 62-3; 2003*b*: 47-8).

Translating the biological knowledge of the cultural other into the categories and theories of global science has arguably been the mission of ethnobiology since the 1950s. This has been so not only in the sense of measuring and comparing local constructs against the Linnean grid, but also in terms of attempting to comprehend local conceptual systems as they pertain to understanding the natural world. Increasingly, that mission has also provided a voice for local people, both technical experts and ordinary folk, as in the role described for indigenous knowledge generally in current development work (Sillitoe, this volume). But since the knowledge which this reflects is articulated orally, or even devolved in non-linguistically coded tacit experience, it often poses major problems for effective conversion into the literate mode, inviting serious oversimplification and straining the limits of ordinary language. Consider, for example, how you would explain to a child how to tie a shoelace – over the telephone. Ethnobiology connects, therefore, with the writing of ethnography,

Until his death in 1988, Ralph Bulmer spent an enormous amount of time in the field working with Kalam subjects in both Papua New Guinea and New Zealand. It is perhaps significant that increasingly he shifted to the publication of verbatim texts, mainly authored by his principal informant, Saem Majnep (Majnep & Bulmer 1977; 1990), which moved him away from an interest in the structure of classifications. In so doing he produced what is perhaps the first postmodern ethnobiology. In contemporary ethnobiology we are more aware of the individual conveyor of knowledge (the Ton Alonsos and the Saem Majneps) than ever before (B. Berlin 2003; Marcus 1991), and it is now difficult to see knowledge simply as disembodied abstract diagrams on paper. In this volume Eugene Hunn explores the notion of 'writing ethnobiology', and provocatively suggests that ethnobiology was producing what James Clifford might call the dialogic ethnographic narrative that allows 'the "subaltern" voice of the Other to be heard' almost a decade before the celebrated volume edited by Clifford and George Marcus (1986). Ethnobiology – like other branches of anthropology – has contributed powerful prose and gifted writing, and has been a prime site for pioneering experimental ethnography and for experimental writing, for example in its innovative use of illustration. Hunn suggests that it is useful to distinguish between popular master narratives, designed to convince the reader (and to his selection we might add, for example, Nabhan 1998), technical narratives, which legitimate the argument with reference to good science, and reflexive monographic narratives, which explore and interpret the data in a self-critical and nuanced way. Amongst the authors of master narratives are some ethnobiologists, such as Richard Schultes (Davis 1996) or Darrell Posey, the charismatic founder of the International Society of Ethnobiology, who are cast in a truly heroic role (see also Plotkin 1993). Such writing is a two-edged sword, for while it advertises the significance of what ethnobiologists do (and may recruit us students and bring us fame, if not fortune), and while there is a clear link between writing, rhetoric, and a kind of advocacy which is central to much ethnobiological research (e.g. Hunn 1990; Posey 1999), it is in danger (like other kinds of popular writing in anthropology) of trivializing and exoticizing.

Cognition and culture

In cognitive anthropology issues raised by ethnobiological classification have been central to debates about category formation, classification, knowledge transmission, and the evolution of the mind. The works of Brent Berlin (1992), Atran (1990), Brown (1984), Hunn (1977), and Boster (1996) all bear testimony to the fertile synergy between research in ethnobiology and cognitive studies more generally. There are important 'inner connections', as Clifford Geertz (1966) might put it, between, say, Berlin's work with Kay (B. Berlin & Kay 1969) on colour, or Conklin's (1964) work on kinship categories, and their work on ethnobiological classification. Ethnobiology has also provided a crucial empirical link connecting anthropology with psychology and cognitive science (Atran 1998), and, in the work of Steve Mithen (1996), with archaeology as well. In this volume, Berlin discusses the striking role of verbal mimesis in the evolution of human cognition, especially in relationship to the semantically opaque names and the physical qualities of certain organisms. Mithen, in his chapter, reiterates the idea that cross-cultural similarities in ethnobiological classification are a legacy of a universal 'evolved predisposition' in *Homo sapiens*. He suggests that a new ability to cognize the natural world accounts, in part, for the success of early species of *Homo*, such as *ergaster*, during the period of Old World colonization about two million years BP, and *Homo sapiens* in the Arctic and the Americas later, and also as some kind of pre-adaptation to the transitions to agriculture of the early Holocene. Mithen presents us with a challenging account of how natural history intelligence might have evolved. Indeed, there can be little doubt that approaches from our understanding of ethnobiological systems will, over the next decade, help to provide answers concerning the extent to which knowledge is constrained by the evolved architecture of the brain, the extent to which that architecture is itself the product of neural enculturation in the development of each biological individual (where the similarities reflect common ecological stimuli), and the extent to which shared knowledge structures reflect parallel socially devolved cultural processes (Ellen 2006).

Many anthropologists reserve their scrutiny for categories which divide up social and cultural space at several removes from unsocialized perception, with so-called 'complex' categories and more abstract schemes. Though such schemes are no less real for those in whose representations they feature, by themselves they serve only as a partial basis for understanding category formation. Ethnobiology has provided much of the evidence through which to test and elaborate theories of the category, in both cognitive anthropology and cognitive psychology. These began by being mainly linguistic in character, and early attempts derive almost entirely from the distinctive feature model developed in componential analysis. Such a model emphasizes the boundaries of categories, and this general approach found favour in the underlying semantics of the Anglo-structuralist analyses of Edmund Leach (1964) and Mary Douglas (1966). What such theories lacked was dedicated research in natural and laboratory settings, and it soon became clear that categories are much more fuzzy in their construction, something which could be modelled by modifying componential analysis as a kind of polythesis, or in terms of core-periphery models which assume the pre-eminence of cognitive prototypes, depending on whether linguistics or psychology is the preferred reference discipline (Ellen 2006: 4-5). At the same time, more attention was being paid to the relationship between cognition as a mental activity and the learned bodily routines which act on and in the world but are not necessarily simply the enaction of mental processes, and between knowledge and enskilment (Ellen 2003a: 48). In other words, ethnobiological ethnography enabled more accurate modelling of real-world categories.

A concern with theories of folk classification, in the sense of attempting to work out the relationship between categories at all degrees of inclusiveness in a domain, is relatively recent, its formative phase being associated largely with Conklin (1954) and Berlin (B. Berlin, Breedlove & Raven 1974). To some extent Émile Durkheim and Marcel Mauss (1963 [1901-2]) had prefigured a sociological theory, but this had never directly addressed mundane or technical categories, though it was adopted by Leach and Douglas as if this had been so. The main impetus, rather, came from work influenced by ethnoscience, adopting the idiom of taxonomy borrowed from the Linnean tradition of Western thought. As with the linguistic paradigm, the taxonomic model proved to be an elegant device for generating data, though its usefulness as an exclusive paradigm has been widely contested (Ellen 1993; Friedberg 1990; Hunn 1977; Sillitoe 2003), in particular its assumptions concerning rank, level, and contrast, together with its weak engagement with the issue of knowledge variation, flexibility in the use of constructs, and the social context of classifying behaviour.

Theories of the change and evolution of lexical and semantic fields relating to environmental sense data are a relatively recent focus of interest (B. Berlin 1972; Brown 1984). We can see an anthropological precursor in Morgan's (1871) theory of the evolution of kinship terminologies. In immediate terms the stimulus can be traced to the aforementioned early work of Berlin and Paul Kay (1969) on the evolution of basic colour terms. Such theories seek to show the order in which terms and ranks are added to languages when the evidence is aggregated at a global level. Clearly, these theories owe little to ethnographic practice; indeed it has been argued that their limitations lie in an over-dependency on inadequate cross-language data, a method of aggregate abstraction which eliminates reference to culture history, the a priori establishment of 'levels', arbitrary assumptions in distinguishing terms as members or otherwise of a particular and exclusive domain, and the inference of psychological reality from nomenclature. However, one lasting impact of this body of work has been the irrefutable conclusion that all cultures encode a concept of basic category, and that they repeatedly divide up the natural world in particular and similar ways.

The social organization and transmission of knowledge

In cognitive anthropology ethnobiology has provided studies that have enlivened and empirically substantiated the more arid debates about knowledge and its transmission. By contrast, social theories have played a relatively peripheral and relatively late role. Durkheim and Mauss gave us a model for the study of social classification, and as we have seen, this was adopted by the Anglo-structuralists and social constructionists. The first sociologically oriented work of any significance was that of Ralph Bulmer (e.g. 1967; 1970), taking his cue from Lévi-Strauss but working very firmly in the British tradition of social anthropology. What is important about Bulmer's work, apart from an impressive thoroughness, is its dedication to dissecting the relationship between the mundane and the symbolic, between what Berlin describes as special-purpose classifications and general-purpose ones. Meanwhile, in North America, Whorfian semantics and the Nida-Conklin hypothesis were instrumental in drawing attention to the complexity of certain terminologies by explaining why we should find certain phylogenetic biases in the distribution of folk knowledge (B. Berlin, Breedlove & Raven 1966).

The last few decades have seen various analyses of the connections between cognition and collective representations, mind and culture, and between 'mundane' and 'symbolic' classifications (e.g. Bloch 1977; 1985; Ellen 2003a: 50-1; 2005). Much of this, including debates around the role of metaphor, totemism, animism, and the construction of 'nature', has supported the view that the interrelationships between symbolic and mundane are often far from clear (Ellen 1993; Fox 1971; Healey 1993; Rival 1998; Rosaldo 1972). Having said as much, some confusion has arisen from a failure to distinguish clearly instruments (means or agents) of cognitive process from the medium of belief and cultural representation, that is, to explore how the invariant possibilities of mind (concretization, polarization, analogy, taxonomy even) act on and through existing sets of beliefs and representations, and influence the formation of new ones. A generation of anthropologists have tended to conflate cognition with collective representations, but as Bloch (1985: 301) has insisted, we cannot treat cognition as some arbitrarily imposed scheme. The kinds of cognitive device which I have just briefly listed are apparent in the social construction of categories across the complete range of human experience, 'mundane' no less than 'symbolic'; and if we wish to understand the processes which underlie classifying activity in general, and which connect the instruments of cognition with ethnographic appearances, we would do well to begin with those processes through which we categorize (as far as this is ever possible) the discontinuities of the physical environment. Despite feedbacks from pre-existing classifications and representations, and their inextricable social contextualization, animals and plants provide us with some of the simplest possible, in a word 'elementary', relationships between objects and their representations that are accessible to researchers producing data in natural settings. Categories of natural kinds are about as rooted in the empirical world as categories can ever be, and in a way that those applied to the world of people and social phenomena can never be.

The kind of work typified by Bulmer, Leach, Douglas, and Bloch, while firmly part of British social anthropology, still did not address empirically the way in which ethnobiological knowledge is distributed, organized, transmitted, or valued. The corrective to the kind of generalization upon which such analyses relied arose, initially, as a critique of the 'omniscient speaker-hearer' model that the ethnoscience exponents had articulated. Increasingly, ethnographic practice began actually to measure the variable distribution of knowledge within a population (e.g. Gardner 1976; Hays 1974), or variation in the significance of particular species (Stoffle, Evans & Olmsted 1990; Turner 1988). But once it became empirically evident that fundamental knowledge might vary within a population, the data raised important issues concerning: the extent of 'cultural consensus' (Ellen 2003a; Romney, Weller & Batchelder 1986; Sillitoe 2003: 109-16); constraints on transmission of knowledge networks deriving from structured bias and stochasticity (Casagrande 2002); knowledge exchange and flow; and the information upon which subsistence decision-making might be based. Here again ethnobiological knowledge provided convenient data with which to explore a new methodology (Boster 1984; 1986). And although much had been assumed and asserted about gender as a variable in knowledge distribution, it was not really until the appearance of the crucial work of Howard (2003) on women and plants that this became a serious matter for empirical investigation. Again, the social divide in knowledge between specialists (e.g. healers) and generalists had much been speculated about, but though crucial to addressing issues of cultural consensus, it has even now proved intractable to effective empirical study. Much more attention has been paid in recent years to knowledge transmission, and here too, much of the impetus has arisen from fears of ethnobiological knowledge erosion. One outcome of a focus on transmission, naturally, was data on the distribution of knowledge by age and generation (Stross 1973). Models for analysing transmission remain an area for current debate central to current anthropological concerns, a debate in which the influence of Luigi Cavalli-Sforza (Hewlett & Cavalli-Sforza 1986; Ohmagari & Berkes 1997) looms large.

Medical ethnobiology
One field where the social organization and distribution of knowledge is well reported is medical anthropology. But although we would like to acknowledge here that medical anthropology has gained much from work conducted by medical ethnobiologists (and vice versa), the case is not a clear one. This is a source of some puzzlement. Textbooks in medical anthropology, on the whole,

pay scant attention to the work of medical ethnobiology. We might reasonably expect that medical ethnobotany or ethnopharmacology or ethnopharmacy would be providing the baseline data for much of the content of medical ethnographies, but what medical anthropology has seemed hitherto to lack is full engagement with phytomedical reality, and an acceptance that the healthcare practices of most people on the planet depend on plants and animals. At the same time many accounts of folk phytomedicinal uses still lack serious consideration of local ethnographic context. Here, it seems to us, is an enormous opportunity and challenge for research.

Of course, there are significant exceptions, the work of Nina Etkin (1986; 1988; 1993) perhaps being the most visible, but there is also the path-breaking work of Elois Ann and Brent Berlin (1996) and (in France) Francis Zimmerman (1989). Work in medical ethnobiology has not only permitted the interrogation of the food/medicine (Etkin 1986; 1994) and medicine/poison (Bisset 1995) conceptual divides, but also the exploration of the interface between the great scholarly systems of knowledge and local folk practice, in, for example, Ayurvedic, Tibetan, and Chinese medicine (Anderson, Salick, Moseley & Ou Xiaokun 2005; Hsu 1999; Zimmerman 1989). The organization of medical and biological knowledge in these traditions is subtly different from much folk medicine, given its literary presentation and institutional base – what Anna Waldstein and Cameron Adams call here 'medical schools'. Its study, therefore, requires different methodological tools and skills from researchers, but it is no less relevant to the broader mission of medical anthropology.

The contribution by Waldstein and Adams to this volume addresses directly the interface between medical anthropology and ethnobiology. It is written as a review – almost an annotated bibliography – precisely because it seeks to build intellectual bridges between medical anthropology and ethnobiology, and precisely because such an infrastructural service to the sub-discipline is what is needed at this time. For there is a curious lack of connection not only between studies of the socio-cultural presentation, treatment, and context of disease and studies of the use of *materia medica* (though see, e.g., Telban 1988), but also between studies of conceptions of the body and bodily experience and studies of the empirical knowledge of anatomy and physiology, excepting, for example, Frake (1980) and Lewis (1974). Coincidentally, work in this second area has been much informed by the same cognitive and empirical approaches which gave rise to the systematic analysis of ethnobiological knowledge and naturalistic ethnomedicine in the 1960s. Moreover, people's folk knowledge of human bodies, their processes and pathologies is closely allied to their experience and understanding of the same dimensions of the animals with which they come into contact, especially if they are domesticated. It is no wonder, therefore, that there is a significant overlap in the herbal treatments reported for humans and those found in the ethnoveterinary literature (Mathias-Mundy & McCorkle 1995). Similarly, just as different medical systems often, and increasingly, coexist in the same place and, indeed, are combined in the therapies of some practitioners and in the minds of consumers (e.g. Golomb 1985), the same has happened historically in terms of the movement of medicinal plants and the ways in which they have been subject to processes of ethnopharmacalogical and therapeutic hybridization (Bennett & Prance 2000).

The applied practice of ethnobiology

Applied ethnobiology did not emerge from within the conventional arena of applied anthropology at all. In one sense it had always been integral to how ethnobotanists at least have conceived their project, as the study of useful (meaning economic) plants. It also existed in a forensic sense, as the specialized techniques by which culturally modified and transformed organic materials can be identified, for whatever applied purpose. As *ethno*-biology rather than economic biology, and specifically as the study of the knowledge of local peoples, it was much rejuvenated in the 1970s through the failure of science-driven top-down development projects, and through the activism of environmental NGOs and indigenous peoples' movements. Because such issues have become intensely political, their 'applied' character has been at times controversial. Its original sense, if we consider for example the practice of 'economic botany' developed at Kew Gardens (Brockway 1979), or what is reflected in the early editorial policy of *Economic Botany*, was the application of biological knowledge to commerce and industry. This is still important, and has its highest, and most controversial, profile in the role of ethnobotany in drug discovery (Chadwick & Marsh 1994). However, the crisis in top-down development policy orientated it firmly from the 1970s onwards towards 'valuation studies' (Brush & Stabinsky 1996; Peters, Gentry & Mendelsohn 1989), and to a role in elucidating farmer knowledge (e.g. Cleveland & Soleri 2002; Richards 1985), knowledge of non-domesticated species, and folk medicinal knowledge, all in the interests of participatory approaches to development (Alcorn 1995; Sillitoe, Bicker & Pottier 2002). From the 1980s, it was being routinely employed as a strategy for ensuring favourable biodiversity outcomes and as a way to integrate the interests of local people with wildlife conservation objectives (Cunningham 2001); it also exercised an influence in agricultural development contexts equal to that of farming systems approaches (Sillitoe 1998: 210). Ethnobiology had become absorbed into the rhetoric of 'indigenous knowledge' and 'indigenous rights'. Indeed, the paradigm of ethnoecology (e.g. Brush 1992; Nazarea 1998; Posey 1984) so massively re-configured studies of agriculture and ecology in small-scale societies that in environmental and ecological anthropology it has virtually obscured and replaced other kinds of approaches.

Unlike conventional applied anthropology, applied ethnobiology at its birth was innocent and unencumbered by the angst and dilemmas accumulated by 100 years of history of anthropology as the purported handmaiden of colonialism. Without this baggage it has managed to achieve recognition for the value of traditional ways of life and affirmed the value of local knowledge systems – as Eugene Hunn puts it in his chapter – without being either romantic or patronizing. Not only has ethnobiology made an enormous impact on the development field and the politics of less developed countries over the last decade – as Paul Sillitoe demonstrates in his chapter – it has also – despite the biopiracy controversies – pioneered protocols for the conduct of responsible research and development of natural products in the face of national and commercial interests (Posey 2000), and been at the forefront of initiatives to promote genuinely participatory approaches, and the interests of local and indigenous populations, especially with respect to the ownership of knowledge and genetic resources. In a workshop at which Darrell Posey and I were once present, I had the temerity to suggest that ethnobiologists ought to be 'dispassionate' in the way they interpreted knowledge systems. As those who knew Darrell might have expected, he demurred, strongly, asserting that, on the contrary, we should be passionate. I like to think we were both right. We should be passionate about knowledge rights and about ethnobiology as a proper subject of study, but dispassionate in our methodologies and in the evaluation of our data.

The co-evolutionary paradigm and biocultural synthesis

As the previous section demonstrates, ethnobiology has provided an ideal site for the convergence of academically driven pure research (largely to do with classification and cognition in anthropology and linguistics) and practice-driven issues to do with subsistence regimes, valuation, and the management of natural resources, more usually associated with biology, agriculture, and forestry. But if ever it was strictly true, it is now no longer possible to describe yourself as an ethnobiologist and operate wholly within a biological paradigm, innocent of a distinctive and emergent dynamic at work in human social and cultural systems. This dynamic simultaneously influences and provides a context for biological change, while biology provides the ultimate conditions for social and cultural continuity. Biological and cultural history (which I take to include processes of natural selection, cultural adaptation, and social continuity) both depend on existing ecological and social contexts for their perpetuation, while these contexts themselves define their existence as outcomes of earlier biological and cultural changes (Figure 1). This is reminiscent of Giddens's notion of structuration in the reproduction of social life (see discussion in Ellen 2003b), and of Ingold's (1986) interpretation of the emergent

16 ROY ELLEN

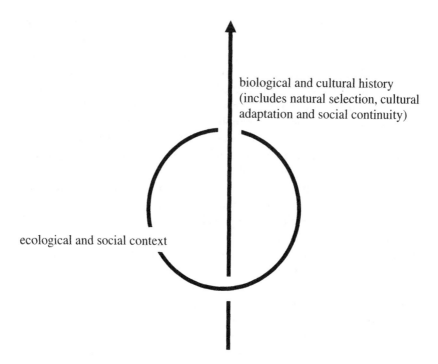

Figure 1. The implicate and recursive relationship between process and context which drives biocultural synthesis.

dynamic between biological and social levels of organization. What is developing as one of the main guiding principles of ethnobiology, therefore, is what we might call the biocultural synthesis, and the failure to see that this has been a core global dynamic for over 10,000 years is an indication of the damage that disciplinary boundaries can do. How could we have ever thought that socio-cultural and biological domains were not inextricably linked, mutually embedded, implicate in each other, if not in simple unicausal relations of determinism?

The key explanatory concept underlying the biocultural synthesis, and one actively fostered by ethnobiological approaches, has been co-evolution. This is reflected, at the highest level of abstraction, in models of gene-culture co-evolution: specifically in work on ethnobiological classification and the idea that cognition of the natural world evolves through interaction with the knowing subject (Boster 1996; Fukui 1996; Mithen 1996). Thus, in what it tells us about the development of human sound symbolism in relation to visual cues in the environment, Berlin provides us here with a strong echo of the relevance of the co-evolutionary concept for understanding synaesthesia. Similarly, we see the influence and power of the co-evolutionary idea in work on the interactions between domestication, nutrition, and the evolution of the human dietary system (Rindos 1984), and on the cultural predisposition to use certain plants in a way we describe as medicinal (Johns 1990; Moerman, Pemberton, Kiefer & Berlin 1999), as discussed here by Waldstein and Adams; in work on forest use and traditional agriculture (Dove 1994; Padoch, Harwell & Susanto 1998); and in work on the history of landscape which traces intricate temporal patterns of interaction between environmental components and cultural practices, or what we now call historical ecology (Balée 1994; Rival in this volume). Historical ecology, with its reliance on ethnobiological data, has become, with ethnoecology, one of the dominant paradigms of the new environmental and ecological anthropology, and is sufficiently robust to also serve as a vehicle for analyses of germplasm exchange and diffusion (e.g. Crosby 1972; Lebot 1991). Similarly, the concept of co-evolution is proving to have wider policy impacts in terms of the way we think about processes of global development in general (Norgaard 1994). But not only have classificatory studies been a primary tool yielding evidence for reconstructing anthropically induced ecological change, historical ecology has provided the context for understanding ethnobiological classification.

Given its emphasis on long-term change, it is hardly surprising that co-evolution has also become a key concept linking the contemporary paradigms of ecological and environmental anthropology to studies of palaeoenvironments and early human development (Crumley 1994; Harris 1989). More so in North America than elsewhere, ethnobiology as an organizing framework has influenced work in archaeology, in the form of archaeobotany (or palaeoethnobotany) and zooarchaeology (e.g. Ford 2001; Willey 1995). While it is true that ethnobiological insights, as with other ethnographic analogies, have for a long time played an important role in interpreting past subsistence systems and ecologies (e.g. Harris 1969; 1977; and in this volume), the conceptual apparatus of historical ecology provides a new and particularly sensitive and methodologically explicit way in which to understand how ecosystem dynamics and human agency intertwine. Indeed, historical ecology and co-evolutionary approaches provide the theoretical context in which Harris's work on multiple alternative pathways to agriculture makes most sense, together with evidence from different regions of the world that challenge the one route to farming approach. Of course, the conceptual and practical difficulties of linking the ethnographic present with the prehistoric past are not to be underestimated. Laura Rival, for example, describes in her chapter the difficulties of assessing the archaeological data from the Amazon, and Harris more generally notes the disparity between the picture emerging from ethnography and that constituted by archaeology. Correlating ethnographic and archaeological evidence presents severe limitations, and Harris meticulously, and in a characteristically measured review, illustrates the strengths and weaknesses in relation to work he has conducted on food procurement using both kinds of data. This disparity is reflected at the methodological and organizational levels of work on the ethnobiology of living peoples and in palaeoethnobiology, the latter often strong on empirical techniques (Hastorf & Popper 1988; Pearsall 1989; Renfrew 1991) but weak on the theoretical linkage it permits with the cultural organization of biological knowledge more generally. Ethnobiology and historical ecology offer new ways in which archaeology and ethnography, and, more pertinently perhaps, archaeologists and ethnographers, can be brought together.

In her contribution to this volume, Rival demonstrates the theoretical power of historical ecology in the context of human ecology more generally. Following Balée (1998: 4), she defines it as the conduct of diachronic analysis of living ecological systems in order to account fully for their structural and functional properties, such that it becomes a 'dialectic of an inalienable link between nature and culture', re-conceptualizing the 'problematic distinction between the wild and the domesticated' so evident in the Western cultural tradition. Though she critiques Balée's concept of post-contact cultural regression, she illustrates powerfully how botanical and ethnobotanical data have come to play an important role in Amazonian anthropology in particular, how tropical forest is being reinterpreted as a mosaic of anthropogenic as well as independent ecological forces (e.g. Ellen in press), and how hunting and gathering in tropical forests therefore need to be re-conceptualized as forms of biological resource management (Posey 1984; Sillitoe 2003). Rival argues that peoples such as the Huaorani 'are active agents in the concentration of useful species', and more. Ethnobiology, therefore, becomes an intrinsic part of historical ecology, which in turn undermines the credibility of the more simplistic, adaptionist, and functional models of human ecology.

The centrality of ethnobiology to anthropology

In anthropology, Wolf (1980), Ortner (1984), and Hart (1990: 14) have all argued that there is no longer a shared discourse, that we are 'over-fragmented, over-specialized', or, to put it more strongly, that we have abandoned the study of humankind (Ingold 1985: 15). This is a widely shared view and one that is hardly new (Needham 1970). Nevertheless, it merits much closer attention than it has hitherto received, especially given its strange inverted echo in Ford's (1978) estimation of the position of ethnobiology in the 1970s with which this introductory essay began. In one sense there never has been a completely shared discourse in anthropology, only a series of overlapping ones; but there is now a resurgent demand within the subject that anthropology concern itself once again with the 'big' issues: with the relationship between naturalism and humanism, with the renegotiation of relations between biology, social life, and culture (Ingold 1985: 15). Ethnobiology, given its subject matter, if not necessarily the practice of individual exponents, can be said to stand at an important intersection between nature, culture, and sociality. For Ford, thirty years on, ethnobiology is now 'at a *crossroads*' (2001: 1, emphasis added). That, I suppose, is some measure of the progress which has been made. I would put it even more strongly: that we have negotiated the crossroads, and are now travelling along a road which has routinized the objectives of ethnobiological study and which continues to yield at an accelerated rate immensely rich and diverse insights which address core anthropological questions. I have attempted here to show how ethnobiology has acquired this conjunctural character, but the message is spelled out in detail and most effectively in the essays which follow.

While, as these essays demonstrate, ethnobiological knowledge is unavoidably affected by cultural relativities, it provides us with a convenient benchmark when examining the perception and use of the environment more widely. Because ethnobiology so obviously focuses on the simultaneity of physical experience, category, and sociality through language, and because it constitutes a domain where the articulation of collective representations with cognitive process, of belief with thought, material with mental, is at its most accessible, it seems to me as good a place as any to begin an inquiry into what we should understand anthropological theory properly to be. Its implications for the study of subsistence behaviour, health, ecology, categorization, and belief may be its most obvious relevancy, but it has also, like kinship in the 1960s (Ortner 1984), provided a convenient body of data on which to experiment using methods involving explicit rigour and formalism, where the search for universals and the idiographic mode engage with rare clarity, and where the relationship between middle-range theory, ethnographic special pleading, and meta-theory is well demonstrated. I think its interdisciplinarity has proved a key to its methodological strength. Interdisciplinary encounters challenge us to make sense of the methods of juxtaposed disciplines, and in ethnobiology the biologists have had to accept the necessity of qualitative methods and of a more critical approach to social and cultural data (Etkin 1993), just as anthropologists have had to develop serious quantitative protocols and measurements to meet the methodological expectations of biologists, and the latter's insistence on sound basic utilitarian procedures, such as voucher specimens, natural resource inventories, and ecological techniques (e.g. Alexiades 1996; Fowler 2001; Martin 1995; Sillitoe 1996; Vogl, Vogl-Lukasser & Puri 2004). Thus, we should not be surprised that ethnobiology has made significant contributions towards the statistical study of cultural consensus, of knowledge variation and transmission. It is methodologically nodal and provides an instructive arena in which to explore the unresolved contradictions inherent in an unusually wide range of theories and orientations. In other words, it is no peripheral frippery but is rather at the heart of that problematic encompassed and scrutinized through the indiscipline of 'anthropology'.

NOTES

[1] I am using the term 'ethnobiology' as shorthand to include ethnobotany, ethnozoology, and ethnoecology, together with any subsidiary designations, such as 'ethnomycology'. Of these, 'ethnobotany' has undoubtedly the greatest presence and intellectual coherence as a subject, but research in any of these fields cannot really proceed far without the other, and in folk knowledge systems there is, anyway, so much empirical overlap and interconnection that it hardly makes sense to distinguish among them. Consider, for example, ethnoveterinary studies, studies of plant insecticides, ethnoentomology, pollination, plant disease, and the analysis of human agricultural and extractive systems and diet. It is for this reason that Darrell Posey was increasingly drawn to 'ethnoecology' as a more encompassing and appropriate term for the kind of work that he undertook.

[2] An inspection of back issues of the journal *Economic Botany* between, say, 1960 and 2004 demonstrates a striking shift towards a focus on ethnobotany and its associated anthropological underpinnings and procedures.

[3] This introduction is to a subject which is accompanied by a large and exponentially growing bibliography. The references cited are, therefore, of selected, indicative and exemplary work only.

REFERENCES

ALCORN, J.B. 1995. The scope and aims of ethnobotany in a developing world. In *Ethnobotany: evolution of a discipline* (eds) R.E. Schultes & S. von Reis, 3-39 London: Chapman & Hall.

ALEXIADES, M.N. 1996. *Selected guidelines for ethnobotanical research: a field manual*. New York: The New York Botanical Garden.

――― 2003. Ethnobotany in the third millennium: expectations and unresolved issues. *Delpinoa* 45, 15-28.

ANDERSON, D., J. SALICK, R.K. MOSELEY & OU XIAOKUN 2005. Conserving the sacred medicine mountains: a vegetation analysis of Tibetan sacred sites in Northwest Yunnan. *Biodiversity and Conservation* (available on-line: http://www.springerlink.com.chain.kent.ac.uk).

ATRAN, S. 1990. *Cognitive foundations of natural history: toward an anthropology of science*. Cambridge: University Press.

――― 1998. Folk biology and the anthropology of science: cognitive universals and cultural particulars. *Behavioural and Brain Sciences* 21, 547-609.

BALÉE, W. 1994. *Footprints of the forest. Ka'apor ethnobotany: the historical ecology of plant utilization by an Amazonian people*. New York: Columbia University Press.

――― (ed.) 1998. *Advances in historical ecology*. New York: Columbia University Press.

BENNETT, B.C. & G.T. PRANCE 2000. Introduced plants in the indigenous pharmacopoeia of northern South America. *Economic Botany* 54, 90-102.

BERLIN, B. 1972. Speculations on the growth of ethnobotanical nomenclature. *Language in Society* 1, 151-86.

―――― 1992. *Ethnobiological classification: principles of categorization of plants and animals in traditional societies*. Princeton: University Press.

―――― 2003. How a folk botanical system can be both natural and comprehensive: one Maya Indian's view of the plant world. In *Nature knowledge: ethnoscience, cognition, and utility* (eds) G. Sanga & G. Ortalli, 38-46. Oxford: Berg.

――――, D. BREEDLOVE & P. RAVEN 1966. Folk taxonomies and biological classification. *Science* **154**, 273-5.

――――, ―――― & ―――― 1974. *Principles of Tzeltal plant classification: an introduction to the botanical ethnography of a Mayan-speaking people of the highland Chiapas*. New York: Academic Press.

―――― & P. KAY 1969. *Basic color terms*. Berkeley: University of California Press.

BERLIN, E.A. & B. BERLIN 1996. *Medical ethnobiology of the Highland Maya of Chiapas, Mexico*. Princeton: University Press.

BISSET, N. 1995. Arrow poisons and their role in the development of medicinal agents. In *Ethnobotany: evolution of a discipline* (eds) R.E. Schultes & S. von Reis, 289-302. London: Chapman & Hall.

BLOCH, M. 1977. The past and the present in the present. *Man* (N.S.) **12**, 278-92.

―――― 1985. From cognition to ideology. In *Power and knowledge: anthropological and sociological approaches* (ed.) R. Fardon, 21-48. Edinburgh: Scottish Academic Press.

BOSTER, J. 1984. Inferring decision making from behaviour: an analysis of Aguaruna Jivaro manioc selection. *Human Ecology* **12**, 347-58.

―――― 1986. Exchange of varieties and information between Aguaruna manioc cultivators. *American Anthropologist* **88**, 429-36.

―――― 1996. Human cognition as a product and agent of evolution. In *Redefining nature: ecology, culture and domestication* (eds) R.F. Ellen & K. Fukui, 269-89. Oxford: Berg.

BROCKWAY, L. 1979. Science and colonial expansion: the role of the British Royal Botanic Gardens. *American Ethnologist* **6**, 449-65.

BROWN, C.H. 1984. *Language and living things: uniformities in folk classification and naming*. New Brunswick, N.J.: Rutgers University Press.

BRUSH, S.B. 1992. Ethnoecology, biodiversity and modernization in Andean potato agriculture. *Journal of Ethnobiology* **12**, 161-85.

―――― & D. STABINSKY 1996. *Valuing local knowledge*. Washington, D.C.: Island Press.

BULMER, R.N.H. 1967. Why is the cassowary not a bird? A problem of zoological taxonomy among the Karam of the New Guinea highlands. *Man* (N.S.) **2**, 5-25.

―――― 1970. Which came first, the chicken or the egg-head? In *Échanges et communications: mélanges offerts à Claude Lévi-Strauss* (eds) J. Pouillon & P. Maranda, 1069-91. The Hague: Mouton.

―――― 1974. Memoirs of a small game hunter: on the track of unknown animal categories in New Guinea. *Journal d'Agriculture Tropicale et de Botanique Appliqué* **21**, 79-99.

―――― 1975. Folk biology in the New Guinea highlands. *Social Science Information* **13**: 45, 928.

CASAGRANDE, D.G. 2002. Ecology, cognition, and cultural transmission of Tzeltal Maya medicinal plant knowledge. Ph.D. dissertation, Department of Anthropology, University of Georgia.

CASTETTER, E.F. 1944. The domain of ethnobiology. *The American Naturalist* **78**: 774, 158-70.

CHADWICK, D.J. & J. MARSH (eds) 1994. *Ethnobotany and the search for new drugs*. (Ciba Foundation Symposium **185**). Chichester: John Wiley.

CLÉMENT, D. 1998a. L'ethnobiologie/Ethnobiology. *Anthropologica* **40**, 7-34.

―――― 1998b. The historical foundations of ethnobiology. *Journal of Ethnobiology* **18**, 161-87.

CLEVELAND, D.A. & D. SOLERI 2002. Introduction: farmers, scientists and plant breeding: knowledge, practice and the possibilities for collaboration. In *Farmers, scientists and plant breeding* (eds) D.A. Cleveland & D. Soleri, 1-18. New York: CAB International.

CLIFFORD, J. & G.E. MARCUS (eds) 1986. *Writing culture: the poetics and politics of ethnography*. Berkeley: University of California Press.

CONKLIN, H.C. 1954. The relation of Hanunóo culture to the plant world. Ph.D. dissertation, Yale University.

——— 1964. Ethnogenealogical method. In *Explorations in cultural anthropology* (ed.) W.H. Goodenough, 25-55. New York: McGraw-Hill.

COTTON, C.M. 1996. *Ethnobotany: principles and applications*. London: John Wiley.

CROSBY, A.W. 1972. *The Colombian exchange: biological and cultural consequences of 1492*. Westport, Conn.: Greenwood Press.

CRUMLEY, C.L. (ed.) 1994. *Historical ecology: cultural knowledge and changing landscapes*. Santa Fe: School of American Research Press.

CUNNINGHAM, A.B. 2001. *Applied ethnobotany: people, wild plant use and conservation*. London: Earthscan.

DAVIS, W. 1995. Ethnobotany: an old practice, a new discipline. In *Ethnobotany: evolution of a discipline* (eds) R.E. Schultes & S. von Reis, 40-52. London: Chapman & Hall.

——— 1996. *One river: explorations and discoveries in the Amazon rain forest*. New York: Simon & Schuster.

DOUGLAS, M. 1966. *Purity and danger*. London: Routledge & Kegan Paul.

DOVE, M.R. 1994. Transition from native forest rubbers to *Hevea brasiliensis* (Euphorbiaceae) among smallholders in Borneo. *Economic Botany* **48**, 382-96.

DURKHEIM, É. & M. MAUSS 1963 [1901-2]. *Primitive classification*. London: Cohen & West.

ELLEN, R.F. 1986. Ethnobiology, cognition and the structure of prehension: some general theoretical notes. *Journal of Ethnobiology* **6**, 83-98.

——— 1993. *The cultural relations of classification: an analysis of Nuaulu animal categories from central Seram*. Cambridge: University Press.

——— 1996. Putting plants in their place: anthropological approaches to understanding the ethnobotanical knowledge of rainforest populations. In *Tropical rainforest research – current issues* (eds) D.S. Edwards, W.E. Booth & S.C. Choy, 457-65. Dordrecht: Kluwer.

——— 2003*a*. Variation and uniformity in the construction of biological knowledge across cultures. In *Nature across cultures: views of nature and the environment in non-Western cultures* (ed.) H. Selin, 47-74. Dordrecht: Kluwer.

——— 2003*b*. Arbitrariness and necessity in ethnobiological classification: notes on some persisting issues. In *Nature knowledge: ethnoscience, cognition and utility* (eds) G. Sanga & G. Ortalli, 47-56. Oxford: Berghahn.

——— 2006. *The categorical impulse: essays in the anthropology of classifying behaviour*. Oxford: Berghahn.

——— in press. Plots, typologies and ethnoecology: local and scientific understandings of forest diversity on Seram. In *Global vs local science* (ed.) P. Sillitoe. Oxford: Berghahn.

——— & H. HARRIS 2000. Introduction. In *Indigenous environmental knowledge and its transformations: critical anthropological perspectives* (eds) R.F. Ellen, P. Parkes & A. Bicker, 1-33. Amsterdam: Harwood.

ETKIN, N.L. 1986. *Plants in indigenous medicine and diet: biobehavioural approaches*. New York: Gordon & Breach.

——— 1988. Ethnopharmacology: behavioural approaches in the study of indigenous medicines. *Annual Review of Anthropology* **17**, 23-42.

—— 1993. Anthropological methods in ethnopharmacology, *Journal of Ethnopharmacology* 38, 93-104.
—— 1994. *Eating on the wild side: the pharmacologic, ecologic and social implications of using non cultigens.* Tucson: University of Arizona Press.
FELD, S. 1996. A poetics of place: ecological and aesthetic co-evolution in a Papua New Guinea rainforest community. In *Redefining nature: ecology, culture and domestication* (eds) R.F. Ellen & K. Fukui, 61-88. Oxford: Berg.
FORD, R.I. 1978. Ethnobotany: historical diversity and synthesis. In *The nature and status of ethnobotany* (ed.) R.I. Ford, 33-50. (Anthropological Papers 67, Museum of Anthropology, University of Michigan). Ann Arbor: University of Michigan.
—— 2001. Introduction: ethnobiology at the crossroads. In *Ethnobiology at the millennium: past promise and future prospects* (ed.) R.I. Ford, 1-10. (Anthropological Papers 91, Museum of Anthropology, University of Michigan). Ann Arbor: University of Michigan.
FOWLER, C.S. 2001. In the field with people, plants and animals: a look at methods. In *Ethnobiology at the millennium: past promise and future prospects* (ed.) R.I. Ford, 149-62. (Anthropological Papers 91, Museum of Anthropology, University of Michigan). Ann Arbor: University of Michigan.
FOX, J. 1971. Sister's child as plant: metaphors in an idiom of consanguinity. In *Rethinking kinship and marriage* (ed.) R. Needham, 219-52. London: Tavistock.
FRAKE, C.O. 1980. *Language and cultural description: essays by Charles O. Frake* (ed.) A.S. Dil. Stanford: University Press.
FRIEDBERG, C. 1990. *Le savoir botanique des Bunaq: percevoir et classer dans le Haut Lamaknen (Timor, Indonésie).* (Mémoires du Muséum National d'Histoire Naturelle, Botanique 32). Paris: Musée National d'Historie Naturelle.
FUKUI, K. 1996. Co-evolution between humans and domesticates: the cultural selection of animal coat colour diversity amongst the Bodi. In *Redefining nature: ecology, culture and domestication* (ed.) R.F. Ellen & K. Fukui, 319-85. Oxford: Berg.
GARDNER, P. 1976. Birds, words and a requiem for the omniscient informant. *American Ethnologist* 8, 446-68.
GEERTZ, C. 1966. Religion as a cultural system. In *Anthropological approaches to the study of religion* (ed.) M. Banton, 1-46. (Association of Social Anthropologists Monograph 3). London: Tavistock.
GOLOMB, L. 1985. *An anthropology of curing in multiethnic Thailand.* Urbana: University of Illinois Press.
HARRINGTON, J.P. 1947. Ethnobiology. *Acta Americana* 5, 244-7.
HARRIS, D.R. 1969. Agricultural systems, ecosystems and the origins of agriculture. In *The domestication and exploitation of plants and animals* (eds) P.J. Ucko & G.W. Dimbleby, 3-15. Chicago: Aldine.
—— 1977. Alternative pathways towards agriculture. In *Origins of agriculture* (ed.) C.A. Reed, 179-243. The Hague: Mouton.
—— 1989. An evolutionary continuum of people-plant interaction. In *Foraging and farming: the evolution of plant exploitation* (eds) D.R. Harris & G.C. Hillman, 11-26. London, Unwin Hyman.
HART, K. 1990. Swimming into the current. *Times Higher Education Supplement* 18 May, 13-14.
HASTORF, C.A. & S. POPPER (eds) 1988. *Current paleoethnobotany: analytical methods and cultural interpretations of archaeological plant remains.* Chicago: University of Chicago Press.

HAUDRICOURT, A. 1973. Botanical nomenclature and its translation. In *Changing perspectives in the history of science: Essays in honour of Joseph Needham* (eds) M. Teich & R. Young, 265-73. London: Heinemann.

——— & L. HÉDIN 1987 [1943]. *L'Homme et les plantes cultivées*. Paris: A.M. Métailié.

HAYS, T.E. 1974. Mauna: explorations in Ndumba ethnobotany. Ph.D. dissertation, University of Washington.

HEALEY, C. 1993. Folk taxonomy and mythology of birds of paradise in the New Guinea highlands. *Ethnology* **32**, 19-34.

HESSE-BIBER, S., P.R. DUPUIS & T. SCOTT KINDER 1997. Anthropology: New developments in video ethnography and visual sociology – analyzing multimedia data qualitatively. *Social Science Computer Review* **15**, 5-12.

HEWLETT, B.S. & L.L. CAVALLI-SFORZA 1986. Cultural transmission among Aka pygmies. *American Anthropologist* **88**, 922-34.

HOLLINGSWORTH, P., R. BATEMAN & R. GORNALL 1999. *Molecular systematics and plant evolution*. (Systematics Association special volume **57**). Basingstoke: Taylor & Francis.

HOWARD, P. (ed.) 2003. *Women and plants: gender relations in biodiversity management and conservation*. London: Zed Press; New York: St Martin's Press.

HSU, E. 1999. *The transmission of Chinese medicine*. Cambridge: University Press.

HUNN, E. 1977. *Tzeltal folk zoology: the classification of discontinuities in nature*. New York: Academic Press.

——— 1990. *Nch'i-wána, 'The Big River': Mid-Columbia Indians and their land*. Seattle: University of Washington Press.

INGOLD, T. 1985. Who studies humanity? The scope of anthropology. *Anthropology Today* **1**: **6**, 15-16.

——— 1986. *Evolution and social life*. Cambridge: University Press.

JOHANNES, R.E. (ed.) 1989. *Traditional ecological knowledge*. Cambridge: IUCN, The World Conservation Union.

JOHNS, T. 1990. *With bitter herbs they shall eat*. Tucson: Arizona University Press.

LAIRD, S.A. (ed.) 2002. *Biodiversity and traditional knowledge: equitable partnerships in practice*. London: Earthscan.

LEACH, E. 1964. Anthropological aspects of language: animal categories and verbal abuse. In *New directions in the study of language* (ed.) E.H. Lenneberg, 23-63. Cambridge, Mass.: MIT Press.

LEBOT, V. 1991. Kava (*Piper methysticum* Forst.f.): the Polynesian dispersal of an Oceanian plant. In *Islands, plants, and Polynesians: an introduction to Polynesian ethnobotany* (eds) P.A. Cox & S.A. Banack, 169-202. Portland, Oreg.: Dioscorides Press.

LÉVI-STRAUSS, C. 1966. *The savage mind*. London: Weidenfeld & Nicolson.

LEWIS, G. 1974. Gnau anatomy and vocabulary for illness. *Oceania* **45**, 50-78.

MAFFI, L. (ed.) 2001. *On biocultural diversity: linking language, knowledge and the environment*. Washington, D.C.: Smithsonian Institution Press.

MAJNEP, I.S. & R. BULMER 1977. *Birds of my Kalam country*. Auckland: Auckland University Press/Oxford University Press.

——— & ——— 1990. *Kalam hunting traditions*, 85-90. (Working Papers in Anthropology, Archaeology, Linguistics, Maori Studies **85-90**). University of Auckland, Department of Anthropology.

MARCUS, G.E. 1991. Notes and quotes concerning the further collaboration of Ian Saem Majnep and Ralph Bulmer: Saem becomes a writer. In *Man and a half: essays in Pacific anthropology and ethnobiology in honour of Ralph Bulmer* (ed.) A. Pawley, 37-45. (Memoir **48**). Auckland: The Polynesian Society.

MARTIN, G.J. 1995. *Ethnobotany: a methods manual.* London: Chapman & Hall.
——— 2002. Ethnobotany, ethnobiology and ethnoecology. *Encyclopedia of Biodiversity* **2**, 609-21.
MATHIAS-MUNDY, E. & C.M. MCCORKLE 1995. Ethnoveterinary medicine and development – a review of the literature. In *The cultural dimension of development: indigenous knowledge systems* (eds) D.M. Warren, L. Slikkerveer & D. Brokensha, 488-98. London: Intermediate Technology Publications.
MEDIN, D. & S. ATRAN (eds) 1999. *Folkbiology.* Cambridge, Mass.: MIT Press.
MINNIS, P. (ed.) 2000. *Ethnobotany: a reader.* Norman: University of Oklahoma Press.
MITHEN, S, 1996. *The prehistory of the mind: a search for the origins of art, religion and science.* London: Thames & Hudson.
MOERMAN, D.E., R.W. PEMBERTON, D. KIEFER & B. BERLIN 1999. A comparative analysis of five medicinal floras. *Journal of Ethnobiology* **19**, 49-67.
MORGAN, L.H. 1871. *Systems of consanguinity and affinity of the human family.* (Smithsonian Contributions to Knowledge 17). Washington, D.C.: Smithsonian Institution.
NABHAN, G.P. 1998. *Cultures of habitat: on nature, culture, and story.* Washington, D.C.: Counterpoint.
——— & A. REA 1987. Plant domestication and folk-biological change: the upper Piman/Devil's Claw example. *American Anthropologist* **89**, 57-73.
NAZAREA, V.D. 1998. *Cultural memory and biodiversity.* Tucson: University of Arizona Press.
NEEDHAM, R. 1970. The future of social anthropology: disintegration or metamorphosis? In *Anniversary contributions to anthropology: twelve essays,* 34-47. Leiden: E.J. Brill.
NORGAARD, R.B. 1994. *Development betrayed: the end of progress and a coevolutionary revisioning of the future.* London: Routledge.
OHMAGARI, K. & F. BERKES 1997. Transmission of indigenous knowledge and bush skills among the western James Bay Cree women of subarctic Canada. *Human Ecology* **25: 2**, 197-222.
ORTNER, S. 1984. Theory in anthropology since the sixties. *Comparative Studies in Society and History* **26**, 126-66.
PADOCH, C., E. HARWELL & A. SUSANTO 1998. Swidden, sawah, and in-between: agricultural transformation in Borneo. *Human Ecology* **26**, 3-20.
PEARSALL, D.M. 1989. *Paleoethnobotany: a handbook of procedures.* San Diego: Academic Press.
PENNISI, E. 2001. Linnaeus' last stand. *Science* **291**, 2304-7.
PETERS, C.M., A.H. GENTRY & R.O. MENDELSOHN 1989. Valuation of an Amazonian rainforest. *Nature* **339**, 655-6.
PLOTKIN, M.J. 1993. *Tales of a shaman's apprentice: an ethnobotanist searches for new medicines in the Amazon rain forest.* New York: Viking.
——— 1995. The importance of ethnobotany for tropical forest conservation. In *Ethnobotany: evolution of a discipline* (eds) R.E. Schultes & S. von Reis, 147-56. London: Chapman & Hall.
PORTERES, R. 1977. Ethnobotanique. *Encyclopaedia Universalis Organum* **17**, 326-30.
POSEY, D. 1984. Ethnoecology as applied anthropology in Amazonian development. *Human Organization* **43**, 95-107.
——— (ed.) 1999. *Cultural and spiritual values of biodiversity.* London: Intermediate Technology Publications.
——— 2000. Ethnobiology and ethnoecology in the context of national laws and international agreements affecting indigenous and local knowledge, traditional resources and intellectual property. In *Indigenous environmental knowledge and its transformations: critical anthropological perspectives* (eds) R.F. Ellen, P. Parkes & A. Bicker, 35-54. London: Harwood.
RELETHFORD, J.H. 2003. *The human species: an introduction to biological anthropology.* Boston: McGraw Hill.

RENFREW, J. 1991. *New light on early farming: recent developments in palaeoethnobotany*. Edinburgh: University Press.
RICHARDS, P. 1985. *Indigenous agricultural revolution: ecology and food production in west Africa*. London: Hutchinson.
RINDOS, D. 1984. *The origins of agriculture: an evolutionary perspective*. Orlando, Fla.: Academic Press.
RIVAL, L. (ed.) 1998. *The social life of trees*. Oxford: Berg.
ROMNEY, K., S. WELLER & W. BATCHELDER 1986. Culture as consensus: a theory of culture and informant accuracy. *American Anthropologist* **88**, 313-38.
ROSALDO, M.Z. 1972. Metaphors and folk classification. *Southwestern Journal of Anthropology* **28**, 83-99.
SCHULTES, R.E. & S. VON REIS (eds) 1995. *Ethnobotany: evolution of a discipline*. London: Chapman & Hall.
SILLITOE, P. 1988. *Made in Niugini: technology in the highlands of Papua New Guinea*. London: British Museum Publications.
——— 1996. *A place against time: land and environment in the Papua New Guinea highlands*. Amsterdam: Harwood.
——— 1998. What, know natives? Local knowledge in development. *Social Anthropology* **6**, 203-20.
——— 2003. *Managing animals in New Guinea: preying the game in the highlands*. (Studies in Environmental Anthropology 7). London: Routledge.
———, A. BICKER & J. POTTIER 2002. *Participating in development: approaches to indigenous knowledge*. (ASA Monographs **39**). London: Routledge.
STEARN, W.T. 1973. *Botanical Latin: history, grammar, syntax, terminology and vocabulary*. Newton Abbot: David & Charles.
STEVENSON, M.C. 1915. Ethnobotany of the Zuni Indians. Bureau of American Ethnology Annual Report 30, 31102. Washington, D.C.: Government Printing Office.
STOFFLE, R.W., M.J. EVANS & J.E. OLMSTED 1990. Calculating the cultural significance of American Indian plants: Paiute and Shoshone ethnobotany at Yucca Mountain, Nevada. *American Anthropologist* **92**, 416-32.
STOLLER, P. 1989. *The taste of ethnographic things: the senses in anthropology*. Philadelphia: Pennsylvania University Press.
STROSS, B. 1973. Acquisition of botanical terminology by Tzeltal children. In *Meaning in Mayan languages* (ed.) M.S. Edmonson, 107-41. The Hague, Mouton.
STURTEVANT, W.C. 1964. Studies in ethnoscience. *American Anthropologist* **66**, 99-131.
TAYLOR, P.M. 1990. *The folk biology of the Tobelo people: a study in folk classification*. (Smithsonian Contributions to Anthropology **34**). Washington, D.C.: Smithsonian Institution Press.
TELBAN, B. 1988. The role of medical ethnobotany in ethnomedicine: a New Guinea example. *Journal of Ethnobiology* **8**, 149-69.
TURNER, N.J. 1988. 'The importance of a rose': evaluating the cultural significance of plants in Thompson and Lillooet Interior Salish. *American Anthropologist* **90**, 272-90.
TUXILL, J. & G.P. NABHAN 2001. *People, plants and protected area*. London: Earthscan.
VOGL, C.R., B. VOGL-LUKASSER & R.K. PURI 2004. Tools and methods for data collection in ethnobotanical studies of homegardens. *Field Methods* **16**, 285-306.
WARREN, D.M., L. SLIKKERVEER & D. BROKENSHA (eds) 1995. *The cultural dimension of development: indigenous knowledge systems*. London: Intermediate Technology Publications.
WHISTLER, W.A. 1991. Polynesian plant introductions. In *Islands, plants, and Polynesians: an introduction to Polynesian ethnobotany* (eds) P.A. Cox & S.A. Banack, 41-66. Portland, Oreg.: Dioscorides Press.

WILLEY, G.R. 1995. Archaeobotany: scope and significance. In *Ethnobotany: evolution of a discipline* (eds) R.E. Schultes & S. von Reis, 400-5. London: Chapman & Hall.
WOLF, E.R. 1980. They divide and subdivide and call it anthropology. *New York Times*, 30 November.
ZENT, S. n.d. A genealogy of scientific perspectives of indigenous knowledge. Draft paper for *Innovative wisdom: the contribution of local knowledge to science* (ed.) G.J. Martin.
ZERNER, C. (ed.) 2000. *People, plants and justice: the politics of nature conservation*. New York: Columbia University Press.
ZIMMERMAN, F. 1989. *Le discours des remèdes au pays des épices*. Paris: Payot.

2

The First Congress of Ethnozoological Nomenclature

BRENT BERLIN

> By establishing a harmony between a thing and its name, we conform to psychic habit as old as humanity (Vendryès 1951 [1925])

> I take for granted, then, that there are some similarities between the experiences we have through different sense organs [and] that in primitive languages one finds much evidence for assuming that the names of things and events often originate according to this similarity between their properties in vision or touch, and certain sounds or acoustical wholes[1] (Köhler 1929: 242)

I have given my chapter the somewhat through-the-looking-glass title 'The First Congress of Ethnozoological Nomenclature'. It is appropriate, then, to begin with part of the conversation between Alice and the Gnat that I included as the epigraph in *Ethnobiological classification* (Berlin 1992: i):

> 'What's the use of their having names,' the Gnat said, 'if they won't answer to them?' 'No use to *them*,' said Alice, 'but it's useful to the people that name them, I suppose. If not, why do they have names at all?' (Carroll 1960: 222).

Given all that we have learned about the relationship of classification and nomenclature since Alice's time, I think we can agree with her that names for creatures must be useful 'to the people that name them'. Much less understood, however, is just *how* and *why* people decide on the names that they bestow on the living creatures that surround them.

So, join me in a fanciful, but perhaps not so implausible, 'First Congress of Ethnozoological Nomenclature', where we listen in on part of the discussion of its delegates in one of their first workshops. The scene opens on a group of sages squatting around an evening campfire somewhere engaged in a serious discussion on a fundamental nomenclatural question, one asked by Roger Brown now more than half a century ago: *how shall a thing be called?*[2]

As we come within earshot of our circle of naming specialists, one of the elders is arguing: 'Then let us assume that if our goal is to propose a name that is a good mnemonic for the referent to which it is assigned, surely an unambiguous descriptive phrase seems ideal'. 'Of those with the enormous bills', he continues, 'the words "the one with underparts, wings and tail black, and upper and lower tail coverts yellow and crimson" are certain to distinguish it from all others of its class' (Fig. 1a).

'Quite so', agrees the elder to his right, 'but such a principle seems hardly appropriate for the inconspicuous grey one that is so common to the forest edges of our garden plots. It seems to call out its name every time it speaks'.

Figure 1. The two birds discussed at the First Congress of Ethnozoological Nomenclature ((a) *Ramphastos vitellinus*; (b) *Lipaugus vociferans*. Reproduced by permission of John A. Gwyne (a) and Guy Tudor (b) in Hilty 2003).

'Of course you refer to the one that says: *uweá, uweá, páipainch.áaa – páipainch.áaa – uweá, uweá, páipainch.áaa – páipainch.áaa.* There seems little doubt but that *páipainch*[3] should be its name' (Fig. 1b). A general nodding of heads all around indicates unanimous assent.

'And your principle would apply equally well to the slippery-slick-skinned ones, *suákarep, karákarás, chirirí, pokarí, and warétete*, that call their names in the night', adds Uncle Skinny as he puts another branch on the fire.

'This works well enough when they *lead* us to their names by their *calls*', pipes up one of the younger members in the circle, a nomenclatural rebel of sorts but who was savvy enough to get invited due to his political connections. 'But what of those that don't call their names and whose descriptive designations are simply ponderous and boring?' Then, looking as if he had discovered the key question, he asks: 'If I were to propose *wámpang* [morpho butterfly] and *wichíkip* [small, inconspicuous butterfly] as the names for two of the silent flutter-flat-wings, is there any doubt about which name we should apply to each?' (Fig. 2).

'The young one speaks the truth', responds the convener of the congress. 'None the less', he concludes, 'I must confess that just *why* this should be so is not at all clear to me. For reasons beyond my explanation, it just seems appropriate and easy to name them in this fashion'.

Our specialist in ethnozoological nomenclature makes a good point. We find it relatively simple to provide explanations for animal names based on descriptive phrases such as *red-winged blackbird* or *redheaded woodpecker* and we can easily offer folk etymologies for names such as *catfish, armour head*, or *bigeye*. Likewise, with little conscious effort we can think of good reasons for the use of onomatopoeic bird names such as *whip-poor-will* and *bobwhite*. Much less understood, however, are those animal names motivated by the principles that Hinton, Nichols, and Ohala have referred to as *synaesthetic sound symbolism*, 'the process whereby certain vowels, consonants, and suprasegmentals are chosen to consistently represent visual, tactile, proprioceptive properties of objects, such as size or shape' (1994: 4). Some of these distinctions are set out in Figure 3.

Early studies of phonaesthesia
One of American linguistics' major figures, Edward Sapir, noted that synaesthetic sound symbolism is at work when we focus on 'the expressively symbolic character of sounds quite aside from what the words in which they occur mean in a referential sense' (1929: 225). Roman Jakobson, principal founder of the Prague School of structural linguistics and phonology, provides the beginnings of a psychological explanation for synaesthetic sound symbolism when he writes:

Figure 2. The two kinds of butterfly discussed at the First Congress of Ethnozoological Nomenclature (*wámpang* large butterfly, top; *wichíkip* small butterflies, bottom (Butterfly Utopia n.d., reproduced with permission).

> The intimacy of the connection between the sounds and the meaning of a word gives rise to a desire by speakers to ... complement the signified by a rudimentary image. Owing to the neuropsychological laws of synaesthesia, phonetic oppositions can themselves evoke [sensations] of pointed and rounded, of thin and thick, of light and heavy (Jakobson 1978: 113).[4]

Synaesthetic sound symbolism can be thought of as the cross-modal mapping that unites specific speech sounds and one or more distinct sense modalities (sight, touch, smell, taste). Recently, a group of electronic vocal composers intrigued by the challenge of making the 'voice visible' have

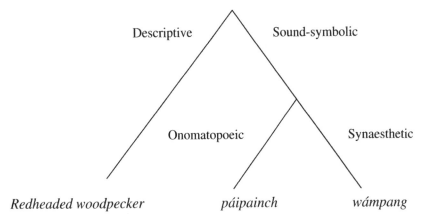

Figure 3. The ethnozoological naming game.

captured the essence of the meaning of synaesthetic sound symbolism in a single term: *phonaesthesia* (Levin, Lieberman, Blonk & La Barbara 2003). This is the term I will use for the sound-symbolic but non-onomatopoeic representation of non-acoustic phenomena.

An overriding feature of the early experimental work on phonaesthesia has been its focus on abstract figures, manufactured objects, nonsense words, and antonyms in natural languages (for some classic early work see Brown 1958a; 1958b; Brown, Black, & Horowitz 1955; Brown & Nuttal 1959). Much less research has been carried out to determine the extent to which phonaesthesia functions as an unconscious psychological motivation that might drive the naming of animals in natural languages. The preliminary data presented here are meant to support the claim that the role of phonaesthesia in ethnozoological vocabulary is greater than any of us might have expected. Furthermore, recent work in neuroscience suggests that phonaesthesia might have played a significant role in the evolution of language, a topic to which I will return at the close of the paper.

However, to place these data in their proper context, let me review two early experiments that set the stage for scores of studies of phonaesthesia for three-quarters of a century. For most American linguistic anthropologists and psycholinguists, Sapir's classic 1929 'A study in phonetic symbolism' paper is probably the best-known experimental work. In one of several studies, Sapir showed his subjects figures of two tables of different sizes, indicating that the nonsense words *mil* and *mal* were the tables' names. He then asked subjects which name indicated the larger table. They invariably chose *mil* for the smaller table and *mal* for the larger (Sapir 1929).

Sapir's research finds its origins in an earlier study by another great linguist, Otto Jesperson, in his little-read paper titled 'Symbolic value of the vowel *i*'[5] (Jesperson 1933 [1922]). Although the data discussed there were anecdotal, they led Jespersen to conclude that in natural languages 'the sound [i] comes to be easily associated with small [things] and [u o a] with bigger things' (1933 [1922]: 284). Common examples of the principle of size-sound symbolism can be seen in the antonyms for 'large' and 'small' in many languages of the world: for example, Spanish *grande – chico*, Tzeltal *muk'ul – ch'in*, Aguaruna *muun – pipich*, and Greek *macros – micros* (for further supporting data on size-symbolism, see Nichols 1971; Sapir 1911; Swadesh 1960; Tsur 1992; 2005; Ultan 1978).

A second experiment, even more widely known than those of Sapir, was carried out by the Georgian psychologist Dimitry Uznadze[6] (1924) and then elaborated five years later by Wolfgang Köhler, the father of Gestalt psychology (1929; 1947). Köhler asked his subjects to observe the two drawings seen in Figure 4 and to assign them the nonsense names *takete* and *maluma* or the basis of what name seemed most appropriate for each figure. More than 90 per cent of the subjects selected *takete* for the angular drawing and *maluma* for the circular one.

Köhler's experiment has been replicated many times (see Fox 1935; Irwin & Newland 1940; Lindauer 1988; 1990 for some early examples), and at least once cross-linguistically among Lilongwe-speaking children of Lake Tanganyika (Davis 1961). I have also informally conducted the experiment regularly with students in my evolution of human cognition classes, always with similar results.

Extending Köhler's experiments on phonaesthesia

Recently, I conceived a modified version of the Köhler study. I hoped to determine what subjects might do if presented with the two figures and asked to

Figure 4. Wolfgang Köhler's (1929: 243) famous figures, *maluma* and *takete*. Reprinted from *Gestalt psychology* by Wolfgang Köhler © Wolfgang Köhler 1929. With permission of the publisher, W.W. Norton & Company, Inc.

produce their *own names* for them? Would the same principles of phonaesthesia that appear to underlie the association of *maluma* and *takete*, with their corresponding figures, be invented *de novo*?[7] Here, briefly, are the details of the study.

Twenty-two students in two introductory philosophy classes at the University of Georgia[8] were requested to be of assistance to Steven Spielberg in his plans to invent a new language called Droidese. The language would be spoken by cartoon characters in a new animated film to be titled *Return of the Droids*. The sound system of Droidese included the consonant and vowel phonemes seen in Table 1.

Students were also informed that a standard rule of Droidese phonology requires that the consonants and vowels of Droidese words take the canonical shape CVCVCV, for example *dawelu*. Given these rules for the form of Droidese words, I then asked them to invent names that they felt would be most appropriate for Köhler's figures (subjects were not told of the *maluma takete* experiment).

The results of the simple study show that vowels and certain consonants of Droidese words exhibit some interesting features (Table 2). The sounds selected to form the invented names for the *maluma* and *takete* figures show significantly different patterns of distribution – *maluma*-like names prefer back vowels [u] and [o] while *takete*-like terms favour the front vowels [i] and [e] (Table 2a).

For consonants, voiced stops were predominantly selected in the formation of words for the *maluma* figure (*takete* terms were about equally divided on the value of voicing) (Table 2b). The patterned distributions of these vowels

Table 1. The sound system of Droidese

consonants					
	p		t		k
	b		d		g
			s		h
	m		n		
			l		
			r		
	w		y		

vowels					
		i		u	
		e		o	
			a		

Table 2. Results of a study of Droidese words.

(a)

Distribution of vowels V1, V2, V3	Maluma-like terms	Takete-like terms	Total
Front vowels i, e	13	30	43
Back vowels u, o	35	16	51
Total	48	46	94

Two-sided p value by Fisher's exact test is .0004.

(b)

Distribution of stop consonants C1, C2, C3	Maluma-like terms	Takete-like terms	Total
Voiced stops	54	36	90
Voiceless stops	12	30	42
Total	66	66	132

Two-sided p value by Fisher's exact test is .0013.

(c)

'Takete' terms	'Maluma' terms
kidise	buromu
tarasi	gobuda
titomi	suhoya
niwabo	woloma
satari	gomobo
keteni	bukona

and consonants were both statistically significant. Examples of some contrasting invented names are seen in Table 2c.

Some typical reasons provided by students to explain their motivations for their invented names were:

> '*Takete*' terms: sounds seemed sharp, straight, diagonal, jagged-edged, rapid, harsh, piercing, pointy, jamming, concise, angular, short, grab attention.
>
> '*Maluma*' terms: sounds seemed to flow smoothly, words sound bubbly, rolling, circular, lazy-sounding vowels, round shape, loose, soft, slow.

One student gave the answer, 'For reasons beyond my explanation, it just seemed appropriate and easy to name them as [I did]', a comment, you'll recall, similar to the one provided by the convener of our First Congress of Ethnozoological Nomenclature.

Students' responses focus on two kinds of visual sense impressions evoked by the sounds of their invented names: one is shape (angular, sharp, curved, jagged-edged, pointy, short, long, diagonal, rigid, twisted); the other is movement (rapid, piercing, jamming, flowing smoothly, bubbly, rolling, fluid, slowly).

These data show that phonaesthesia is associated in significant ways with the sensory sensations of size (Sapir's *mil, mal*), and shape and movement (invented words for *maluma, takete*). While an adequate psycholinguistic explanation for these results has yet to be proposed, recent research has gone far in providing us with partial answers as to why phonaesthesia seems to be so intuitively natural. Building on earlier work of John Ohala (1984), Hinton, Nichols, and Ohala provide the following synthesis:

> [H]igh tones, vowels with high second formants [F_2] (notably /i/), [a vowel's formants are acoustic properties] and high-frequency consonants are associated with high-frequency sounds, small size, sharpness, and rapid movement; low tone, vowels with low second formants [F_2], (notably /u/ [and, we could probably add, /a/ and /o/], and low-frequency consonants are associated with low-frequency sounds, large size, softness, and heavy, slow movements (1994: 10).[9]

These perceptual properties of phonaesthesia can be illustrated in Table 3.[10]

Table 3. Some perceptual properties of phonaesthesia.

	Semantic dimension of phonaesthesia			
Sound segment	Movement	Shape	Size	Sound
Front vowels, high-frequency consonants	Rapid, quick, fast, abrupt, short, agile	Long, straight, jagged, skinny, sharp, thin, slender, angular, lanky, pointed, flat (a special case of long)	Small	High tone
Back vowels, Low-frequency consonants	Slow, sluggish, lumbering, flowing, waddling, awkward, smooth	Spherical, short, stocky, fat, round, solid, hefty, squat, rotund, smooth	Large	Low tone

Phonaesthesia in ethnozoological nomenclature

In what ways might these non-arbitrary sound-meaning associations of *movement*, *size*, and *shape*, so typical of phonaesthesia, be applicable to how animals are named by peoples in traditional societies? In earlier research, I described how size-sound phonaesthesia could be shown to be at work in the ethnozoological nomenclature for birds and fish among the Huambisa, a Jívaro-speaking people of Amazonas, Peru (Berlin 1992; 1994; see also comparable work on Malay fish names, Berlin 2005). Phonaesthesia was prevalent in both lexical domains – smaller birds and fish tended to exhibit names formed with the high-front vowel [i] while larger birds and fish favoured vowels [u] and [a] significantly greater than expected by chance. In addition, but only mentioned in passing in my first reports, it appears that fish and bird names differ significantly in the distribution of high acoustic frequency consonants: bird names overwhelmingly prefer initial voiceless stops [p, t, k] and voiceless affricates [ts, ch] before the high-frequency vowel [i]; when these consonants occur as initial segments in names of fish, they overwhelmingly co-occur with the low-frequency vowels [a] and [u].

In addition, fish names show a distinct preference for nasal consonants, sound segments with extremely low sound frequencies. Thus, relative size can be said to be marked by vowel quality whereas movement appears to be partially marked by acoustic quality of consonants. These data also suggest that a combinatorial effect is achieved when the vowel and consonant parameters act in concert.

Might the names of other groups of animals show a similar sound-symbolic naming bias, and, if so, how widespread might the bias be across other languages? I began to collect data on this question when I was struck by the natural, non-arbitrary nature of the Aguaruna names for the lumbering South American tapir, *pamáu*, and the quick and agile squirrel, *wichíng* (compare *wámpang* and *wichíkip* in Figure 2; see also Berlin 2004).

My loosely formed hypothesis was that if the principles of phonaesthesia were at work, one should expect to find terms for tapir comprised of sounds of low acoustic frequency (large size, slowness), and words for squirrel marked by sounds of high acoustic frequency (small size, quickness). I gathered comparable data from a sample of twenty-five South American Indian languages each from a distinct linguistic family. The results of this comparison are seen in Table 4. The data show that the high-front vowel [i] and the high-frequency voiceless stops [t] and [k] mark terms for squirrel (small size, rapid movement?) while the sounds [i], [t], and [k] are not favoured in the names for tapir (due to their large size and slow movement?).

Are there naming contrasts relating to other creatures that might form the beginnings of a database on which we might base a theoretical explanation of

Table 4. The results of a comparison of high and low acoustic frequency sounds in words for squirrel and tapir in twenty-five languages.

a)

vowel [i]	Squirrel	Tapir	Total
+	19	7	26
−	4	16	20
total	23	23	46
two-sided P value is 0.0008			

b)

vowel [a]	Squirrel	Tapir	Total
+	8	19	27
−	15	4	19
total	23	23	46
$P = 0.0023$			

c)

Stops [t], [k]	Squirrel	Tapir	Total
+	14	5	19
−	9	18	38
total	23	23	46
two-sided P value is 0.0015			

Labials [p], [b], [m], [w]	Squirrel	Tapir	Total
+	9	16	25
−	14	7	21
total	23	23	46

$P = 0.0746$ (not quite significant)

phonaesthesia in ethnozoological nomenclature? Turning again to South America, I selected two groups of birds, the rails and the tinamous, both of which differed significantly in shape (and movement?). Drawings of a species from each group are seen in Figure 5. Visually, these two birds might be thought of as metaphorically analogous to the *takete* and *maluma* figures in Köhler's

Figure 5. *Aramides cajanea* (rail) (a) and *Tinamous major* (tinamou) (b) (reproduced with the permission of Guy Tudor from Meyer de Schauensee & Phelps 1978).

studies. As a thought experiment, imagine that the names of the two creatures in Figure 5 in some indigenous language spoken in the Upper Amazon of Colombia were *maluma* and *takete*. Which name would one assign to each figure? In another experiment, I aimed to hold my own First Congress of Ethnozoological Nomenclature.

Using the instructions developed for Köhler's *maluma* and *takete* figures, I substituted the *takete/maluma* figures with drawings of a rail and a tinamou. My hypothesis was similar to the first study: on the basis of visual clues based on form (and the real or metaphorical association of form with type or speed of movement), names for the rail should show high-frequency consonants and vowels while those chosen for the tinamou would be of low frequency.

I asked sixty-two undergraduate anthropology students,[11] none of whom had participated in the earlier experiments, to invent names for the two birds. I passed around test sheets with the sound system of Droidese accompanied by drawings of each bird. I further explained Droidese word formation as outlined in the first experiment but stated that all words must take the canonical form CVCVC, a highly common consonant-vowel sequence for word formation in many of the world's languages.[12]

When students had finished writing their invented names for each of the birds, I then asked them to watch me write two words on a white board at the

Table 5. Vowel and consonant patterns for invented rail and tinamou names in student experiments.

(a)

Vowels	Rail	Tinamou	Total
i, e	71	36	107
o, u	38	60	98
Total	109	96	205

$p = .0001.$

(b)

Stops	Rail	Tinamou	Total
Voiceless C1, C2, C3	64	38	102
Voiced C1, C2, C3	37	72	109
Total	101	110	211

$p = .0001.$

front of the class, stating that if the words I had written might be assigned to the rail and the tinamou, which name would go with what bird? The names were *taket* and *malum*, variants of Köhler's terms. Students then wrote each of these names next to the bird on their test sheet to which they felt it most appropriately applied.

First, as you would have predicted, sixty-one of the sixty-six students (92 per cent) stated that *taket* was the most appropriate name for the rail. When we look at the invented names for rail and tinamous, we see a familiar pattern: high-front vowels [i] and [e] are preferred for the rail, back vowels [o] and [u] for the tinamou (Table 5a). With consonants, we see that voiceless stops [p], [t], and [k] are favoured in the invented names for rail, much less so for tinamou (Table 5b).

The reasons students gave for their naming responses support the principles of phonaesthesia for shape and movement discussed earlier:

> Rail: fast, long, hard, thin, lanky, light, agile, angular, lean, movement, pointy, quick, sharp, high-strung (taut), skinny.
>
> Tinamou: big, corpulent, dark, thick, fat, waddles, round, husky, heavy, rotund, awkward, low, plump, fullness, stocky, 'daubling' [sic], soft, solid, squat, slow.

What are the naming patterns for rail and tinamou in the languages of the traditional societies who actually know these birds intimately? With the goal of gathering initial data on this question, I focused once more on South America. I soon discovered that gathering lexical data for these bird species proved more difficult (and perhaps less reliable) than my earlier efforts to develop a comparative list of terms for tapir and squirrel. It is common that dictionaries or word lists do not indicate to which species a particular term may apply. When dictionaries do list terms for tinamou and rail, names for tinamous, the most obvious and culturally significant group, are cited more often than terms for rails. Furthermore, a folk name might include several biological species, some of the same scientific genus, some not. The 'true' rail and tinamou species in South America are:

Rails (Rallidae): *Aramides cajanea, A. axillaries, A. erythroptera, Rallus longirostris.*

Tinamous (Tinamidae): *Tinamus major, T. tao, T. guttatus, Crypturellus soui, C. cinerus, C. undulatus, C. varigatus.*

As I believed it important to draw a sample of languages that included both a word for 'rail' and 'tinamou', I have limited my sample to languages where I am fairly confident of the terms listed for the large tinamous, *Tinamus* spp. (mostly commonly, *T. major* and *T. tao*) and species of the Wood Rail, *Aramides* spp. (most commonly, the Grey-Necked Wood Rail, *Aramides cajenea*). This resulted in a sample of seventeen languages of distinct language families, or, when in the same family, showing rail and tinamou terms that are not obvious cognates, as is the case with Sataré and Urubu of the Tupian language family (Table 6).

Conclusions and possible explanations

Even with such a small sample, if phonaesthesia is at work here, the principles noted for shape and movement should be evident. The distributional patterns of high- versus low-frequency vowels, it turns out, are not significant but the predicted patterns for consonant phonaesthesia are suggestive. An angular, sharp, long-legged, streamlined-bodied rail ought to show a preference for voiceless consonants, especially voiceless stops, while the rounded, short-legged tinamou should not favour these sounds. Table 7a shows that high-intensity stops [t] or [k] occur in all terms for rail; [p], the least intense of the stops, occurs in just two names.

In contrast, nasals are consonants with a particularly low acoustic frequency and should connote slow, round, plump, fat, soft, squat, heavy creatures such as the tinamou. These sounds should be avoided in terms for rail and favoured for tinamous (Table 7b).

Table 6. Names for rail and tinamou in seventeen languages.

Language	Language family	Scientific name	Native name
Achuar	Jivaroan	Aramides cajanea Tinamous major	kawachaa wa
Apalai	Cariban	Aramides cajanea Tinamous major	kutuka póhno
Arabela	Zaparoan	Aramides sp. Tinamus sp.	shiitoioru* napanaha
Bora	Witotoan	Aramides sp. Tinamous major	pookoroji aawaa
Capanahua	Panoan	Aramides sp. Tinamus sp.	tako koma
Chayahuita	Cahuapanan	Aramides sp. Tinamus sp.	konsha' shonshoron
Cuiba	Guahiban	Aramides cajanea Tinamus major	kotsato mami
Ese-eje	Takanan	Aramides sp. Tinamus major	ta'jo tobi sha'wa
Jarawara	Arawá	Aramides cajanea Tinamus major	wakari yimo
Kandoshi	Lg. isolate	Aramides sp. Tinamou sp.	píkoróro katamshi
Piaroa	Salivan	Aramides cajanea Tinamus major	mu'kaeni wewa
Mashco-Piro	Arawakan	Aramides sp. Tinamus spp.	kokru kwawa
Quechua	Quechuan	Aramides sp. Tinamus major	Kutsra Yutu
Sateré-Mawé	Tupian	Aramides cajanea Tinamus major	tarangku urit'iwato
Siona	Tucanoan	Aramides cajanea Tinamus tao	bo'te anka
Ulwa	Misumalpan	Aramides cajanea Tinamus major	kudah wankamara
Urubu	Tupian	Aramides cajanea Tinamus major	sarakur inambu

* ii = long vowel.
Language sources: Achuar: Fast, Warkentin & Fast 1996, Apalai: Jensen 1998; Arabela: Rich 1999; Bora: Thiesen & Thiesen 1998; Capanahua: Loos & Loos 2003; Chayahuita: Hart 1988; Cuiba: Ortiz & Queixalos n.d.; Ese-Eje: SAILDP n.d.; Jarawara: Vogel, n.d.; Kandoshi: Tuggy 1966; Piaroa: E. Zent, personal communication; Mashco-Piro: SAILDP n.d.; Quechua: Irvine 1987; Sateré-Mawé: Jensen 1998; Siona: Wheeler 1987; Ulwa: Dictionary of the Ulwa language n.d.; Urubu: Jensen 1998.

Table 7. Consonant phonaesthesia in names for rails and tinamous in seventeen languages.

(a)

[p], [t], [k]	Rail	Tinamou	Total
+	17	10	27
−	0	7	7
Total	17	17	34

$p = .0072$.

(b)

Nasals	Rail	Tinamou	Total
+	3	10	13
−	14	7	21
Total	17	17	34

$p = .0324$.

(c)

Labials	Rail	Tinamou	Total
+	6	14	20
−	11	3	14
Total	17	17	34

$p = .0134$.

Finally, all *labial* consonants, [p], [b], [m], and [w] exhibit low frequency in terms of air flow and as a consequence should be associated with the expected principles of phonaesthesia governing shape and movement. Terms for rail should show significantly fewer labials than do terms for tinamous (Table 7c).

The distributional patterns observed for consonant phonaesthesia in these seventeen languages are strong. It is likely that data from additional languages will confirm the expected principles of shape/movement sound symbolism. This expectation is strengthened given the studies described earlier in this paper.

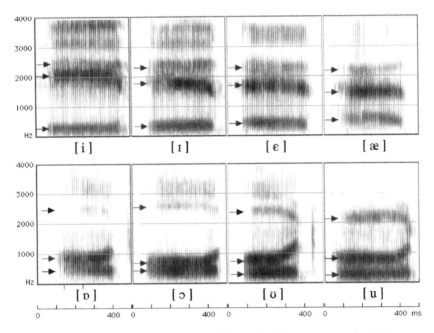

Figure 6. A spectrogram of the words *heed, hid, head, had, hod, hawed, hood, who'd* as spoken by a male speaker of American English. The locations of the first three formats are shown by arrows. (after Ladefoged n.d., reproduced with permission.)

In searching for an explanation of these findings, the data show that the physical properties (relative acoustic frequencies) of vowels and consonants play a major role. The high-front vowel [i] has an acoustic frequency of around 2100-2300 Hz as its second formant (F_2, in acoustic terminology) while the back vowels [u, o] exhibit F_2 between 700 and 800 Hz (Fig. 6).

Physical acoustic parameters are also of relevance in consonant phonaesthesia. Voiceless stops are characterized by an 'explosion' in the normal airflow during articulation since in 'consonants, voiceless obstruents have higher frequency than voiced because of the higher velocity of the airflow, ... dental, alveolar, palatal and front velars [have] higher frequencies (of bursts, frication noise, and/or formant transitions) than labials and back velars' (Ohala 1994: 331).

Furthermore, the voiceless stops [p], [t], and [k] are graded in terms of the intensity of the airflow associated with their production (p < t < k) (Fig. 7). Ohala states it as follows:

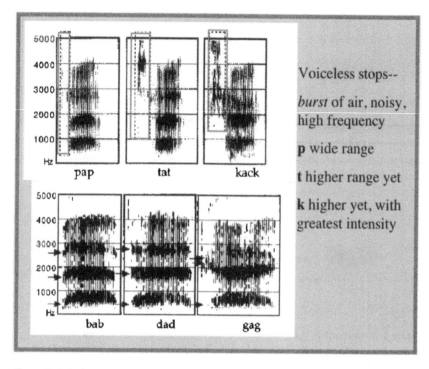

Figure 7. Voiceless stops graded in terms of intensity of airflow (after Fant 1973: 114, reproduced with permission).

> [B]oth [t] and [k] have high-intensity bursts. One factor is that they are inside the vocal tract and thus exploit the downstream resonances. [p] can't do this since there is no downstream cavity. [t] has a shorter cavity than [k] but it has the advantage that the air stream of the release can hit the teeth (lower and upper incisors) and these act as what the aerodynamic people call 'spoilers' or baffles, creating another source of turbulence (Ohala, personal communication; see also Tsur 2001).

A second parameter at play in phonaestheia is the articulation of speech sounds, something that might be referred to as *vocal mimesis*, the unconscious use of mouth gestures to indicate metaphorically the inherent qualities of the thing being named. This process was first systematically discussed by the British speech therapist Sir Richard Paget in his *Human speech* (1930), published just a year after Sapir's sound symbolism study. Paget argued that:

> The sound of the word is frequently found to be due to postures and gestures of the organs of articulation which bear a pantomimic relation to the idea or action to which the word refers. From this we infer that human speech arose out of a generalized unconscious

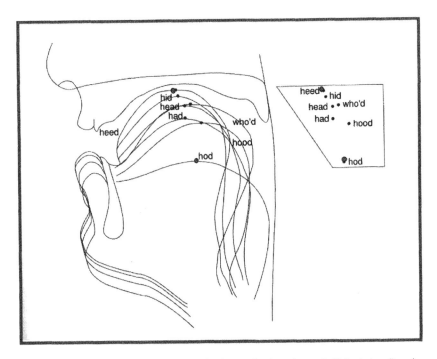

Figure 8. Relative positions of the tongue in the production of sounds [i] (as in heed) and [a] (as in hod) (after Ladefoged 2002: 115, reproduced with permission).

> pantomimic gesture language (including the tongue and lips). The gestures [of speech] were recognized by the hearer because the hearer unconsciously reproduced in his mind the actual gesture which had produced the sound (1930: 174).[13]

Continuing,

> ... [i] is made by pushing the tongue forward and upward so as to make the smallest cavity between the tongue and the lips while o-o or aw-aw [o] and [a] are the results of a lowered tongue, producing a large mouth cavity (1930: 178) (Fig. 8).

Paget's views mesh closely with Merlin Donald's speculations in *Origins of the modern mind: three stages in the evolution of culture and cognition* (Donald 1991; 1997; also see 2001) and the evolutionary speculations of Peter MacNeilage and his colleague B.L. Davis on motor mechanisms in speech production (MacNeilage 1994; 1998; MacNeilage & Davis 2000; 2001). In speculating on the role of mimesis in the development of human language, Donald notes that the principle of self-triggered voluntary retrieval of representations had to be established in the brain before the highly complex motor acts of speech would have been possible. Phonetic skill [making speech possible] depends on the basically mimetic ability of individuals to create rehearsable and retrievable vocal acts, usually in close connection with other mimetic acts. In a word, language ... is layered on top of a mimetically-skilled phonological system (1997: 728).

Donald is at a loss, however, to provide a plausible theory for the development of lexicon in the first place. He states that, granted phonological agility, *'Lexical invention is not yet understood in terms of mechanism. There is no viable computational model of this process'* (1997: 738, emphasis added). None the less, Donald does anticipate sound-symbolic processes in early word formation when he continues: 'Lexical invention must work according to a metaphorical principle' (1997: 741; see also Foster 1992; Hiraga 2000; Lakoff & Johnson 1980). If phonaesthesia is, above all, metaphorical, then sound symbolism would be a good candidate to first drive lexical representation in spoken language.

These speculations also tend to support Mithen's (1996) proposal that the human capacity for metaphorical thought arises from increasingly complex human symbolic processes as a result of what he calls 'cognitive fluidity' across the four principal domains of intuitive knowledge – social, natural history, technological, and linguistic. The cross-modal metaphorical properties of phonaesthesia would serve only to facilitate the cognitive fluidity that Mithen suggests is essential for higher cognitive development (see Mithen 1996, esp. chap. 10, and in this volume). In this light, the phonaesthesia that we find in ethnozoological vocabulary could have actively contributed to the development of humans' natural history intelligence in ways that are supremely intuitive – and natural.

Recently, and unknown to me until this paper had already appeared in draft form, these plausible yet speculative ideas have received some tentative neurological support from the research of cognitive scientists S.V. Ramachandran and E.M. Hubbard (2001). They agree that what we have been talking about is 'a kind of sensory-to-motor synaesthesia which may have played a pivotal role in the evolution of language' (2001: 19). They argue that 'the reason [subjects get the names right in the *maluma/takete* experiment] is that the sharp changes in visual direction of the lines mimics the sharp [phonetic] inflections of the sounds as well as the [mimetic] movements of the tongue on the palate' (2001: 19).

They conclude that 'the representation of ... lip and tongue movements in [our] motor brain maps may be mapped in non-arbitrary ways to certain sound inflections in auditory regions [of the brain] and the latter, in turn, function as non-arbitrary links to an external object's visual appearance' (2001: 20).

We may tentatively conclude that non-arbitrary sound-symbolic, phonomimetic reference must have had enormous adaptive significance for our hominid ancestors as they began to play the naming game in earnest. It is not too great a stretch of the imagination to suggest that the intuitively plausible and metaphorically motivated principles of phonaesthesia served to drive the development of the lexicon in general. The strength of this type of sound symbolism is manifest in our names for the animals that surround us and, we must assume, the very act of suitably naming them played a critical role in our linguistic and cultural evolution.

We continue to recognize the emotive power of linking *creature* and *name-of-creature* in the zoological lexicon of current-day local languages spoken by members of traditional societies. It is likely, as Köhler and others have suggested, that marking natural kinds with verbal labels that reflect their non-arbitrary *inherent properties* is more apparent in the languages of small, local populations than in the modern world languages of technologically complex populations. And it should hardly be news that the focus on form (shape), movement, and relative size is so apparent, given that these three perceptual dimensions mark the most significant inherent properties of living things that humans immediately recognize and aim to represent symbolically in a creature's name (see Adams & Conklin 1973; Clark 1974).

From the conservative perspective of historical linguistics, it is somewhat disconcerting that phonaesthesia remains so strong a nomenclatural principle in spite of the levelling effects of regular sound change and the subsequent loss of sound-meaning associations that accompanied the development of duality of patterning – perhaps one of the most important signposts in language evolution. But good, workable solutions are not easily abandoned – as *pamau* and *wiching* for tapir and squirrel, *tungkau* and *íspik* for giant doradid catfish and little angelfish, and *wa* and *kawachaa* for tinamou and rail so clearly demonstrate. These are as good names as we are likely to find in the naming game and, should they some day be replaced, those that follow will be equally intuitively satisfying, to both our eyes and ears. Thus, by carrying on in the tradition of our colleagues at the First Congress of Ethnozoological Nomenclature, our names for animals continue to mirror the life of the world of nature to which they have been, and will continue to be, symbolically assigned.

NOTES

This chapter owes much to the critical comments of John Ohala. His proposals on the 'frequency code' underlying sound symbolism (Ohala 1984) forms the empirical foundation for all serious research on the topic. I have not always taken his advice and any major errors on acoustic phonetics are my own. Allen Jensen and other researchers of the Summer Institute of Linguistics have provided important original data from their long-term linguistic work with South American Indian languages. This paper could not have been written without access to their resources. I also wish to note my appreciation for the support and encouragement of Roy Ellen. I am grateful to him for his kind invitation to read an earlier version of this paper in his symposium at the Ninth International Congress of Ethnobiology and for his cogent editorial advice in helping me move a rough oral presentation into its current written form. Finally, I thank Elois Ann Berlin, whose useful and ingenious suggestions helped me design the experiments described in the text.

[1] Since the times of Plato's *Cratylus* (360 BC), it has been recognized that names in natural languages should in some way be non-arbitrarily assigned to their referents.

> Then [says Socrates to Hermogenes], I should say that this giving of names can be no such light matter as you fancy ... and Cratylus is right in saying that things have names by nature, and that not every man is an artificer of names, but he only who looks to the name which each thing by nature has, and is able to express the true forms of things in letters and syllables (Plato n.d.).

This position contrasts markedly with the prevailing view in modern linguistics that flows directly from Ferdinand de Saussure's doctrine on the arbitrary relationship of speech sounds and meaning. The strength of de Saussure's views has been so pervasive that serious studies of sound symbolism were relegated to the backwaters of linguistic theory until recently (Ciccotosto 1991; Foster 1992; Hinton, Nichols & Ohala 1994; Linguistic Iconism Association n.d.; Nuckolls 1996; 1999). The empirical research available on the non-arbitrariness of the relation of sound and meaning in language now demonstrates that the standard doctrine is surely inadequate. One is no longer accused of practising fringe linguistics by conducting research on sound symbolism as a pervasive semantic process in language.

[2] My paper is dedicated to the late social psychologist Roger Brown, who published an important paper with this title (1958*b*). His work on the psychology of language, and especially sound symbolism, has continued to inspire me since the time I first read his *Words and things* (1958*a*).

[3] *Páipaich* is the name of *Lipaugus vociferans* in Aguaruna, a Jívaroan language of Amazonas, Peru.

[4] It is appropriate to recognize here the work of Dwight Bolinger and Yakov Malkiel, primary proponents of sound-symbolic studies in mid-twentieth-century American structural linguistics. Bolinger coined the term 'phonosemantics' (1946), whereas Malkiel, focusing primarily on historical problems, preferred the term 'phonosymbolism' (1990). Bolinger credits Tolman (1887, 1904) for having conducted sound symbolism research prior to that of Sapir (Bolinger, personal communication).

[5] This high-front unrounded vowel is written phonetically [i], as pronounced in English *see*, Spanish *mil* 'one thousand', or Japanese *mi* 'fruit, nut'.

[6] In his honour, Professor Uznadze is remembered today by the Dimitry Uznadze Institute of Psychology, Tbilisi, Republic of Georgia.

[7] So far as I have been able to discover from a review of the extensive experimental literature using the Köhler figures – and it is vast – this is the first experiment that required subjects to produce their own names for the drawings.

[8] I appreciate the collaboration of Richard Shedenhelm, Department of Philosophy, for allowing me to use the students in his classes for this experiment.

[9] Note that the three major perceptual dimensions that focus on movement, shape, and relative size are precisely the basic organizational properties of objects that humans acquire in their categorization of the world in general (see Adams & Conklin 1973; Clark 1974).

[10] After producing this summary table, I discovered in a citation by Fox (1935: 448) that the well-known Africanist anthropologist M. Westermann (1927) had published a remarkably similar table.

[11] I thank Professors Mikel Gleason and Mark Williams for allowing me to use students in their introductory classes for this experiment. For the present paper, I report only on the results from the experiment conducted in Professor Gleason's class.

[12] My specific instructions were, 'After looking closely at the birds, under each bird write down a Droidese name that you believe best represents their inherent qualities, names that just seem natural to you'.

[13] Paget's ideas on vocal gestures were apparently inspired by his work with deaf children and his efforts (with Grace Paget and Pierre Gorman) to develop a manual sign language, referred to today as Paget-Gorman Signed Speech.

REFERENCES

ADAMS, K.L. & N.F. CONKLIN 1973. Toward a theory of natural classification. In *Papers from the Ninth Regional Meeting of the Chicago Linguistic Society* (eds) C. Corum, T.C. Smith-Clark & A. Wieser, 1-10.

BERLIN, B. 1992. *Ethnobiological classification: principles of categorization of plants and animals in traditional societies*. Princeton: University Press.

——— 1994. Evidence for pervasive synesthetic sound symbolism in ethnozoological nomenclature. In *Sound symbolism* (eds) L. Hinton, J. Nichols & J. Ohala, 76-93. Cambridge: University Press.

——— 2004. Tapir and squirrel: further nomenclatural meanderings toward a universal sound-symbolic bestiary. In *Nature knowledge: ethnoscience, cognition, and utility* (eds) G. Sanga & G. Ortalli, 119-27. New York: Berghahn Books.

——— 2005. 'Just another fish story?' Size-symbolic properties of fish names. In *Animal names* (eds) A. Minelli, G. Ortalli & G. Sanga, 9-21. Venice: Istituto Veneto di Scienze, Lettere ed Arti.

BOLINGER, D. 1946. The sign is not arbitrary. *Boletin del Instituto Caruso y Caro* **5**, 52-62.

BROWN, R. 1958a. *Words and things*. New York: Free Press.

——— 1958b. How shall a thing be called? *Psychological Review* **65**, 14-21.

———, A.H. BLACK & A.E. HOROWITZ 1955. Phonetic symbolism in natural languages. *Journal of Abnormal and Social Psychology* **50**, 388-93.

——— & R. NUTTAL 1959. Method in phonetic symbolism experiments. *Journal of Abnormal and Social Psychology* **59**, 441-5.

BUTTERFLY UTOPIA n.d. Website (*http://www.butterflyutopia.com/391.html*).

CARROLL, L. 1960. *The annotated Alice: Alice's adventures in wonderland and through the looking glass* (with introduction and notes by M. Gardner). New York: Charkson N. Potter, Inc.

CICCOTOSTO, N. 1991. Sound symbolism in natural language. Ph.D. dissertation, University of Florida, Gainesville, Florida.

CLARK, E.V. 1974. Classifiers and semantic acquisition: universal categories? Paper presented at the 73rd annual meeting of the American Anthropological Association, Mexico City, 19-24 November.

DAVIS, R. 1961. The fitness of names to drawings: a cross-cultural study in Tanganyika. *British Journal of Psychology* **52**, 259-68.

DICTIONARY OF THE ULWA LANGUAGE n.d. Website (*http://www.slaxicon.org/ulwa/dict/*).

DONALD, M. 1991. *Origins of the modern mind: three states in the evolution of culture and cognition*. Cambridge, Mass.: Harvard University Press.

——— 1997. Précis of *Origins of the modern mind: three stages in the evolution of culture and cognition*. *Behavioral and Brain Sciences* **16**, 737-91.

——— 2001. *A mind so rare: the evolution of human consciousness*. New York: W.W. Norton and Company.

FANT, G. 1973. *Speech sounds and features.* Boston: MIT Press.
FAST, G., R. WARKENTIN & D. FAST 1996. *Diccionario Achuar-Shiwiar-Castellano.* (Serie Lingüística Peruana 36). Lima: Ministerio de Educación/Instituto Lingüístico de Verano.
FOSTER, M.L. 1992. Body processes in the evolution of language. In *Giving the body its due* (ed) M. Sheets-Johnstone, 208-30. Albany: State University of New York Press.
FOX, C.W. 1935. An experimental study in naming. *American Journal of Psychology* 47, 545-79.
HART, H. 1988. *Diccionario Chayahuita-Castellano* (Serie Lingüística Peruana 29). Lima: Ministerio de Educación/Instituto Lingüístico de Verano.
HILTY, S.L. 2003. *Birds of Venezuala.* (Second edition). Princeton: University Press.
HINTON, L., J. NICHOLS & J. OHALA (eds) 1994. *Sound symbolism.* Cambridge: University Press.
HIRAGA, M. 2000. The interplay of metaphor and iconicity: a cognitive approach. Unpublished Ph.D. thesis, University of London.
IRVINE, D. 1987. Resource management by the Runa Indians of the Ecuadorian Amazon. Ph.D. dissertation, Stanford University, Stanford, California.
IRWIN, F.W. & E. NEWMAN 1940. A genetic study of the naming of visual figures. *Journal of Psychology* 9, 3-16.
JAKOBOSON, R. 1978. *Six lectures on sound and meaning.* Cambridge, Mass.: MIT Press.
——— & R. WAUGH 1987. *The sound shape of language.* Berlin: Mouton de Gruyter.
JENSEN, A.A. 1998. *Sistemas indígenas de classificação de aves: aspectos comparatives, ecológicos, e evolutivos.* Belém-Pará: Ministério de Ciencia e Tecnologia, Conselho Nacional de Desenvolvimento Científico e Tecnológico, Meseu Paraense Emílio Goeldi.
JESPERSON, O. 1933 [1922]. Symbolic value of the vowel *i*. In *Linguistica*, 283-303. College Park, Md.
KÖHLER, W. 1929. *Gestalt psychology.* New York: Liveright Co.
——— 1947. *Gestalt psychology.* (Second edition). New York: Liveright Co.
LADEFOGED, P. 2002. *Vowels and consonants: an introduction to the sounds of languages.* Oxford: Blackwell.
——— n.d. *A course in phonetics* (available on-line: *http:hctv.humnet.ucla.edu/departments/linguistics/VowelsandConsonants/course/chapter8/8.3htm*).
LAKOFF, G. & M. JOHNSON 1980. *Metaphors we live by.* New York: Barnes and Noble.
LEVIN, G., Z. LIBERMAN, J. BLONK & J. LA BARBARA 2003. *Messa di voce*: an audiovisual performance and installation for voice and interactive media (available online: *http://tmema.org/messa/messa.html#background*).
LINDAUER, M.S. 1988. Size and distance perception of the physiognomic stimulus *taketa*. *Bulletin of the Psychonomic Society* 26: 3, 217-20.
——— 1990. The meanings of the physiognomic stimuli *taketa* and *maluma*. *Bulletin of the Psychonomic Society* 28: 1, 47-50.
Linguistic Iconism Association n.d. Website (*http://www.conknet.com/~mmagnus/LIA/*).
LOOS, E. & B. LOOS 2003. *Diccionario Capanahua-Castellano* (Serie Lingüística Peruana 45). Lima: Ministerio de Educación/Instituto Lingüístico de Verano.
MACNEILAGE, P. 1994. Prolegomena to a theory of the sound pattern of the first spoken language. *Phonetica* 51, 184-94.
——— 1998. The frame/content theory of evolution and speech production. *Behavioral and Brain Sciences* 21, 499-511.
——— & B.L. DAVIS 2000. On the origin of internal structure of word forms. *Science* 288, 527-31.
——— & ——— 2001. Motor mechanisms in speech ontogeny: phylogenetic, neurobiological and linguistic implications. *Current Biology* 11, 696-700.
MALKIEL, Y. 1990. *Diachronic problems in phonosymbolism: edita and inedita, 1979-1988*, vol. 1. London: John Benjamins.

MEYER DE SCHAUENSEE, R. & W.H. PHELPS 1978. *A guide to the birds of Venezuela*. Princeton: University Press.
MITHEN, S. 1996. *The prehistory of the mind: a search for origins of art, religion and science*. London: Thames & Hudson.
NICHOLS, J. 1971. Diminutive consonant symbolism in western North America. *Language* 47, 826-48.
NUCKOLLS, J.B. 1996. *Sounds like life: sound-symbolic grammar, performance, and cognition in Pastaza Quechua*. Oxford: University Press.
——— 1999. The case for sound symbolism. *Annual Review of Anthropology* 28, 225-52.
OHALA, J. 1984. An ethnological perspective on common cross-language utilization of F_0 of voice. *Phonetica* 41, 1-16.
——— 1994. The frequency code underlies the sound-symbolic use of voice pitch. In *Sound symbolism* (eds) L. Hinton, J. Nichols & J. Ohala, 325-47. Cambridge: University Press.
ORTIZ, F. & F. QUEIXALOS n.d. Ornitología Cuiva-Guahibo (available on-line: http://www.vjf.cnrs.fr/celia/FichExt/Am/A_06_06.htm).
PAGET, R. 1930. *Human speech: some observations, experiments, and conclusions as to the nature, origin, purpose, and possible improvement of human speech*. London: Routledge & Kegan Paul.
PLATO n.d. *Cratylus* (trans. B. Jowett; available on-line: http://eserver.org/philosophy/plato/cratylus.txt).
RAMACHANDRAN, V.S. & E.M. HUBBARD 2001. Synaesthesia: A window into perception, thought, and language (available on-line: http://psy.ucsd.edu/chip/pdf/Synaesthesia%20-%20JCS.pdf).
RICH, R. 1999. *Diccionario Arabela*. (Serie Lingüística Peruana 49). Lima: Ministerio de Educación/Instituto Lingüístico de Verano.
SAILDP (South American Indian Languages Documentation Project). n.d. Data files available in personal data files of B. Berlin, Laboratories of Ethnobiology, University of Georgia, Athens, Georgia.
SAPIR, E. 1911. Diminutive and augmentative consonant symbolism in Wishram. In *Handbook of American Indian languages*, Bull. 40, Part I, 638-46. Washington, D.C.: Bureau of American Ethnography.
——— 1929. A study in phonetic symbolism. *Journal of Experimental Psychology* 12, 225-39.
SWADESH, M. 1960. On inter-hemispheric linguistic connections. In *Culture history: essays in honor of Paul Radin* (ed) S. Diamond, 894-924. New York: Columbia University Press.
THIESEN, W. & E. THIESEN 1998. *Diccionario Bora-Castellano-Castellano-Bora*. (Serie Lingüística Peruana 46). Lima: Ministerio de Educación/Instituto Lingüístico de Verano.
TOLMAN, A.H. 1887. The laws of tone color in the English language. *Andover Review* 7, 326-37.
——— 1904. Symbolic value of English sounds. In *The views about Hamlet, and other essays*, A.H. Tolman. New York: Houghton Mifflin.
TSUR, R. 1992. *What makes sound patterns expressive? The poetic mode of speech-perception*. Durham, N.C.: Duke University Press.
——— 2001. Onomatopoeia: cuckoo-language and tick-tocking: the constraints of semiotic systems (available on-line: http://www.trismegistos.com/IconicityInLanguage/Articles/Tsur/default.html).
——— 2005. Sound symbolism revisited (available on-line: http://www.tau.ac.il/~tsurxx/SizeSound/Size-Sound_Symbolism.pdf).
TUGGY, J. 1966. *Vocabulario Candoshi de Loreto*. (Serie Lingüística Peruana 2). Lima: Ministerio de Educación/Instituto Lingüístico de Verano.
ULTAN, R. 1978. Size-sound symbolism. In *Universals in human language*, vol. 2: *Phonology* (ed.) J. Greenberg, 525-68. Stanford: University Press.

UZNADZE, D. 1924. Ein experimentaller Beitrag zum Problem der psychologischen Grundlagen der Namengebung. *Psychologische Forschung* 5, 24-43.
VENDRYÈS, J. 1951 [1925]. *Language: a linguistic introduction to history* (trans. P. Radin). New York: Barnes & Noble.
VOGEL, A. n.d. Dicionário Jarawara-Português. Sociedade Internacional de Lingüística, Cuiabá, MT (available on-line: *http://www.sil.org/americas/brasil/PUBLCNS/DICTGRAM/JADict.pdf*).
WESTERMANN, D. 1927. Laut, Ton und Sinn in westafrikanische Sudansprachen. In *Festschrift Meinhof*, 315-28. Hamburg: Kommissionsverlag von L. Friederichsen & Co.
WHEELER, A. 1987. *Gantëya bain: el pueblo Siona del río Putumayo, Colombia*, vol. II. Lomalinda, Meta Colombia: Editorial Townsend.

3
Ethnobiology and the evolution of the human mind

STEVEN MITHEN

Palaeoanthropologists concerned with human evolution have a 'give and take' relationship with ethnobiology as defined by Roy Ellen (this volume): 'the study of how people of all, and of any, cultural tradition interpret, conceptualize, represent, cope with, utilize, and generally manage their knowledge of those domains of environmental experience which encompass living organisms, and whose scientific study we demarcate as botany, zoology, and ecology'. With regard to 'take', palaeoanthropologists need to draw on the theories and findings of ethnobiology to facilitate the interpretation of animal and plant remains recovered from archaeological sites. Such interpretations require inferences about past decision-making, such as about which animals to hunt or which seeds to sow, and hence issues of environmental perception and classification are paramount. These are, of course, difficult if not impossible for palaeoanthropologists to address in an entirely satisfactory manner. Nevertheless, a minimum requirement is that palaeoanthropologists are conversant with relevant issues in ethnobiology and draw on these – as far as is possible – to make informed interpretations of the animal bones and plant remains that they excavate. Making such informed interpretations is quite different from imposing ill-considered ethnographic analogies onto the past, as David Harris illustrates in his paper in this volume. With regard to 'give', palaeoanthropologists need to inform ethnobiologists about the implications of the fossil and artefactual record for the evolution of human mentality, and hence what type of perceptual, classificatory, and decision-making processes we should expect to exist in the minds and cultures of modern humans today, as it is these which condition the human engagement with the natural world.

My concern in this chapter is with giving rather than taking. After providing a brief résumé of human evolution, I will consider a selection of issues that suggest the likely existence of universal cognitive systems for engaging with the natural world and the significance of non-linguistic approaches in ethnobiology.

The fossil and archaeological record for human evolution

Human cognition shares a great many features with that of not only other primates but also other mammals in general. My concern, however, is with those features that are distinctively human – those that distinguish us from our closest living relative, the chimpanzee. It was between six and eight million years ago when the *Pan* and *Homo* lineages diverged from a common ancestor, with the latter having a single extant species, *H. sapiens*. Although the fossil record becomes richer year by year, and occasionally throws up a surprise such as the discovery of the claimed new species *Homo floresiensis* in 2003 (Brown et al. 2004), the basic outline of human evolution within the *Homo* lineage is unlikely to be substantially altered by new discoveries, especially as it is increasingly supported by evolutionary genetics (Jobling, Hurles & Tyler-Smith 2004).

In essence, there were numerous species of bipedal primate in Africa between six and two million years ago. The fossilized remains display considerable morphological variation, which suggests the past exploitation of a variety of specific niches within the African landscape (Johansen & Edgar 1996; Lewin & Foley 2003). The hominins have been classified according to three genera, *Ardipithecus, Australopithecus*, and *Homo*, with the latter constituted by two purported species, *H. habilis* and *H. rudolfensis*. Flaked stone tools are known from at least 2.5 million years ago (Schick & Toth 1993), but in the light of the repertoire of tools used by chimpanzees (Whiten et al. 1999) it seems likely that hammer stones, sticks, leaves, and other minimally modified materials were used long before flaked stone artefacts appeared.

The two species of early *Homo* prior to 1.8 million years ago are characterized by relatively small teeth and flat faces, along with larger brains than the other hominins, reaching up to 750 cc rather than the 450 cc which is also characteristic of chimpanzees today (Wood 1992). Such brain sizes may be within the range of variation for this grade of hominin without necessarily indicating newly evolved cognitive or linguistic abilities. The key problem we face when assessing the significance of brain size is the rarity of the post-cranial skeletons for the earliest *Homo*, which might indicate that the relatively large brains are simply a product of large body size rather than encephalization. Indeed, a case has been made for reclassifying *H. habilis* and *H. rudolfensis* as australopithecines, and for identifying *H. ergaster*, appearing by 1.8 million years ago, as the first member of the *Homo* genus (Wood & Collard 1999).

H. ergaster may mark an evolutionary transition to a type of hominin for which behavioural analogies with living non-human primates are of limited value. With a near-fully modern stature and bipedal gait, this species is most likely the first to have dispersed out of Africa (Straus & Bar-Yosef 2001). Brain size reached up to 900 cc, although some specimens show the maintenance of relatively small brain capacities – those from the site of Dmanisi in Georgia, for example, are no more than 650 cc (Gabunia *et al.* 2000). The Asian lineage of this species evolved into *Homo erectus* and may have made at least one water crossing, to reach Flores Island by 850,000 years ago (Morwood, O'Sullivan, Aziz & Raza 1998).

In Europe, the middle Pleistocene is marked by a succession of hominins that are claimed to constitute at least three species: *H. antecessor*, *H. heidelbergensis*, and *H. neanderthalensis*. These appear to show relative stability in brain size until *c.*600,000 years ago, after which there was rapid encephalization until a capacity equivalent to, and in some cases exceeding, that of *Homo sapiens* is attained in the late Pleistocene, 1,200-1,750 cc (Ruff, Trinkaus & Holliday 1997). Stone artefacts appear to increase in technological complexity from Oldowan-like flakes associated with *H. antecessor*, Acheulian handaxes with *H. heidelbergensis*, and levallois technology with *H. neanderthalensis*. Examples of bone and wooden tools are exceedingly rare, but the discovery of the Schöningen spears (Thieme 1997) indicates that this is most likely a consequence of preservation and discovery. Traces of structures and non-utilitarian artefacts are effectively absent (Gamble 1999), with the few ambiguous examples that exist merely serving to emphasize their extreme rarity and unsophisticated nature.

The hominin species of Europe may form a single evolving lineage or mutliple dispersals into that continent of species that evolved in Africa. Within Africa there is evolutionary continuity from *H. ergaster* to *H. sapiens* (McBrearty & Brooks 2000), with the earliest specimens of the latter dating to *c.*200,000 years ago (McDougall, Brown & Fleagle 2005; White *et al.* 2003), a date that effectively coincides with an estimate for the origin of *H. sapiens* from the study of modern-day genetic diversity (Ingman, Kaessmann, Paabo & Gyllensten 2000; Jobling, Hurles & Tyler-Smith 2004). While *H. sapiens* specimens in Israel indicate initial dispersal out of Africa around 100,000 years ago (Lahr & Foley 1994), the genetic evidence indicates that it was after 60,000 years ago that major dispersals into Asia and Europe occurred, and that it was these that gave rise to the extant populations today (Ingman, Kaessmann, Paabo & Gyllensten 2000).

Immediately prior to such dispersals we find evidence in South Africa for new types of material culture, often assumed to reflect the appearance of symbolic thought and language. Most notable are the incised ochre nodules and shell beads from Blombos Cave dating to 70,000 years ago (Henshilwood et al. 2002; Henshilwood, d'Errico, Vanhaeren, van Niekerk & Jacobs 2004), while red ochre is prevalent in Middle Stone Age deposits at South African sites reaching back to 100,000 years ago (Knight, Powers & Watts 1995).

The dispersals of *H. sapiens* out of Africa resulted in the colonization of Australia by at least 30,000 years ago, most probably between 40,000 and 45,000 years ago (O'Connell & Allen 2004), and that of Europe by 40,000 years ago (Mellars 2004). The latter is associated with major technological innovations characterized by the Aurignacian culture. The Neanderthals may have attempted to imitate the culture of the incoming *H. sapiens*, although d'Errico and colleagues argue that the Neanderthals independently invented the Upper Palaeolithic-type features of the Chatelperronian industry (d'Errico, Zilhão, Julian, Baffier & Pelegrin, 1998). The Neanderthals were either out-competed for resources or were unable to survive the major climatic fluctuations of the late Pleistocene (d'Errico & Sanchez Goni 2003; Stringer, Barton & Finlayson 2000). By 30,000 years ago, the *H. sapiens* in Europe were engaging in cave painting, carving intricate bone figurines, and making elaborately decorated burials (Gamble 1999). Similarly, the *H. sapiens* in Australia and Africa were likely to be engaged in both abstract and figurative rock art at this date (Mulvaney & Kamminga 1999).

The Neanderthals became extinct soon after 30,000 years ago. Precisely when *H. erectus* in Asia became extinct remains unclear; a similar date to that of the Neanderthals is likely, although the claimed species *H. floresiensis* may have survived on Flores island until as late as 14,000 years ago (Morwood et al. 2004). The climatic deterioration of the last glacial maximum of 20,000 years ago caused *H. sapiens* to abandon northern landscapes that became polar desert, and those areas in low latitudes that became extremely arid. As global warming began at c.15,000 years ago, *H. sapiens* re-colonized these landscapes and proceeded to colonize the rest of the world either in the late Pleistocene or early Holocene – the far north, the Americas, and the islands of the Pacific (Mithen 2003).

It was only in the Holocene, beginning a mere 11,600 years ago, that agricultural economies developed, initially in the Near East at c.9,000 years ago and then quite independently at several locations elsewhere in the world – including rice farming in China at c.7,000 years ago, that of maize and squash in Central America at c.6,000 years ago, and domesticated camelids (llamas and alpacas) in the Peruvian Andes by c.5,000 years ago (Mithen 2003; Smith 1995). Farming provided the economic foundation for the development of towns and within a few thousand years the first 'civilizations', within which writing was independently invented.

It is within this span of human evolution that I have so briefly summarized that the distinctive properties of the human mind evolved. That evolutionary history has not only provided the potential for the astonishing cultural diversity that we find within the world today, and perhaps even more so in the recent past, but also imposes constraints on the character of thought and behaviour. Understanding both the potential and the constraints is crucial for the development of culturally informed biological knowledge and practices, and suggests that some approaches may be more fruitful than others. In this regard, I will now consider a suite of selected issues from the evolutionary history of humankind that appear especially relevant to ethnobiology.

The paradox of genetic unity and cultural diversity
A decade ago palaeoanthropologists were debating whether *Homo sapiens* had a recent African origin or was the product of multi-regional evolution. The accumulation of fossil, archaeological, and genetic evidence has conclusively demonstrated the former to be correct (Jobling, Hurles & Tyler-Smith 2004), although this does not preclude inter-breeding between *Homo sapiens* dispersing from Africa and other species of *Homo* in Eurasia. The recent origin of *Homo sapiens* has emphasized the human paradox of an extraordinary degree of cultural diversity underlain by a relatively limited extent of genetic variation. Human genetic diversity is indeed highly constrained, with significantly greater differences between chimpanzees separated by a few kilometres in Africa than between humans living at the opposite ends of the earth and engaged in quite different lifestyles (Jobling, Hurles & Tyler-Smith 2004).

The need to reconcile genetic similarity and cultural diversity requires a cognitive anthropology, one that is concerned with how observed behaviour is a product of the interaction between universal properties of the human mind and the unique environmental, cultural, and historical settings within which individuals are located (Mithen 1990). Hence ethnobiologists should be immediately biased towards expecting underlying commonalities in the classification systems of the natural world, which will have various cultural manifestations.

Mental modularity and the evolution of the mind
Throughout our evolutionary history, not just that which I have briefly summarized above but that which extends into far more distant times, our ancestors have had to acquire information and make decisions about the natural world for their survival and reproduction. The principles of natural selection suggest that those individuals who were more able to acquire relevant information and process such information in the most effective manner would have gained reproductive success. In many species, the relevant information for

exploiting a very specific niche has simply been encoded within the animal itself through natural selection, so there is effectively no process of learning involved. Our hominin ancestors were, however, generalists and lived in highly variable spatial and temporal environments (deMenocal 1995). Rather than having specific responses to environmental stimuli encoded into their genome, natural selection has provided learning and decision rules to enable behavioural flexibility.

The extent to which such rules are of a general-purpose nature and limited in number – such as associative or conditional learning – or numerous and specialized for particular domains of activity, has been a key topic of debate in evolutionary psychology (Carruthers & Chamberlain 2000; Mithen 1996; Samuels 1998). While the latter has been favoured by those who argue from first principles of natural selection (e.g. Cosmides & Tooby 1994; Pinker 1997), the former appears more in accord with our observances of our closest relative, the chimpanzee, which appears adept at a wide range of activities and able readily to learn new tasks, such as use of computer keyboards (Byrne 1996a). Chimpanzees, however, have relatively small brains with respect to not only modern humans but also the hominins of the Middle and Later Pleistocene. They are, therefore, relatively uninformative about the evolution of a specifically human type of intelligence.

My own view is that some degree of mental modularity arose during the course of human cognitive evolution that included learning and decision rules tailored by natural selection to interact with the natural world (Mithen 1990; 1996). One key feature of such rules would have been a means of classifying the natural world so as to reduce the information that requires processing when making decisions. Humphrey eloquently summarizes why classification is important:

> In order to be effective agents in the natural world, animals require the guidance of a 'world model', an internal representation of what the world is like and how it works. This model enables them to predict in advance the characteristics of 'recognisable' objects, to anticipate the likely course of events in the environment, and to plan their behaviour accordingly. The role of classification in this context is to help organise sensory experience and to introduce an essential economy into the description of the world. An effective classification system is one which divides up objects in the world into discrete categories according to criteria which make an object's membership of any particular class a relevant datum for guiding behaviour: the objects in any one class may differ in detail but they should share some essential features which give them a common significance for the animal. Such a classification system will reduce the 'thought load' on the animal, expedite new learning, and allow rapid and efficient extrapolation from one set of circumstances to another (1984: 126-7).

A critical time period for the evolution of classificatory systems and other mental modules relating to the natural world in *Homo* is the Plio-Pleistocene, when the archaeological record indicates significant increases in meat-eating and tool production (Isaac 1983; 1989; Potts 1988). These behavioural developments are associated with increased brain size, changing dentition, and bipedalism (Aiello 1996; Aiello & Wheeler 1995; Wheeler 1991). Identifying cause and effect is difficult, but such anatomical and behavioural developments appear to arise as a package with strong feedback loops and are ultimately related to the change from a forest to a woodland savannah landscape. They appear to culminate in the appearance of *Homo ergaster*, which some anthropologists would designate as the first member of our genus (Wood & Collard 1999).

'Social brain' or 'ecological brain'?

Dunbar (1996; 2004) has argued that the expansion of the brain during not only the Plio-Pleistocene but the whole of human evolution relates to the development of a specifically social intelligence, characterizing this theory as the 'social brain hypothesis'. His claim is based on a correlation between group size and brain size (as measured by neo-cortex ratio) among extant primates and the absence of any clear association between brain size and foraging behaviour (Aiello & Dunbar 1993; Dunbar 1993). Measures of social complexity in terms of instances of 'deceptive' behaviour also suggest a correlation with brain size (Byrne 1996b) and hence may support the view that encephalization is primarily related to the development of a social intelligence, although this does not necessarily rule out general-purpose abilities.

The key feature of social intelligence that Dunbar's work promotes is that of 'theory of mind' – the ability to understand that other individuals may have beliefs and desires that are different to one's own, although the definitions and cognitive interpretations of 'theory of mind' are complex (Carruthers & Smith 1996). One characterization of theory of mind is in terms of 'orders of intentionality': if I know what I think, then I am termed as having a single order of intentionality; if I know what someone else thinks, then I have two orders of intentionality; if I know what someone else thinks that a third partly thinks, then I have a third order of intentionality – and so forth. Whereas humans routinely use three or four orders of intentionality in their social life, apes might be limited to two orders at most (Dunbar 2004).

Whether chimpanzees have such cognitive abilities remains unclear in spite of a vast quantity of theoretical debate, experimental studies, and field observations devoted or contributing to this issue (e.g. Byrne & Whiten 1988; 1992; Povinelli 1993; 1999). The latest research suggests that some features of theory of mind/multiple orders of intentionality may well be present in the chimpanzee mind (Tomasello, Call & Hare 2003). Nevertheless, it is unquestionably the case that these are significantly more evolved within the *Homo* lineage, which partly accounts for the social complexity of our species and is most likely related to the evolution of language (Dunbar 1998; Mithen 1999).

There are, however, three key problems with the 'social brain' hypothesis. First, while a correlation between brain and group size may hold within extant primates that have brain sizes up to $c.400$ cc (although there is some doubt as to whether this is indeed the case – Steele 1996), there is little justification for extrapolating this relationship to the much larger-brained hominins with brain sizes up to 1,750 cc. Even if the initial phase of brain expansion at around two million years ago can be accounted for by the evolution of a social intelligence, there is no reason why the same should apply to the more pronounced period of brain expansion after 600,000 years ago. Second, it seems unlikely that the addition of further levels of intentionality necessarily requires significantly more brain matter: to go from first to second and then to third levels of intentionality simply requires a recursive loop so that the same cognitive mechanism, presumably some type of neural network, is repeatedly used. Third, it is not only in their social relationships that modern humans and extinct hominins are/were quite different to living primates: foraging and tool-making are also considerably more complex and are likely to have required the evolution of additional neural networks and hence the expansion of the brain.

In the light of these weakness with the 'social brain' hypothesis, it seems more plausible to explain the expansion of the human brain as reflecting the development not of a specifically social intelligence but either of multiple intelligences, with each relating to a particular domain of behaviour, or simply of a general-purpose intelligence. The latter appears unlikely in light of the evidence for multiple intelligences/cognitive domains in the modern human mind (e.g. Gardner 1983; 1993; Hirschfeld & Gelman 1994) and the arguments for mental modularity from the principles of natural selection (Cosmides & Tooby 1994; Pinker 1997). In place of a single general-purpose intelligence, I have proposed elsewhere (Mithen 1996) that there were three isolated cognitive domains in the minds of large brain hominins (i.e. post-*Homo habilis*) that I described as social, technical, and natural history intelligences, with each being constituted by a bundle of interacting mental modules.

Our interest here is with natural history intelligence, which might also be characterized as 'intuitive biology'. By this I mean stores of information about the natural world, methods of acquiring further information, and methods of processing information that had become embedded within the hominin genome and did not require learning and/or cultural transmission to acquire. Moreover, these were dedicated to the natural world and quite different to the stores of information and processing methods that were dedicated to interacting with other individuals or manufacturing artefacts. Such intuitive biology would have provided hominins with the type of expert folk-botanical and folk-zoological knowledge that is characteristic of recent hunter-gatherers.

Here I am particularly thinking about their ability to infer information about animals and plants from signs such as tracks, scats, noises, smells, and broken vegetation (e.g. see Gubser 1965: 121, 221-2, 230; Hill & Hawkes 1983; Lee 1979: 212-13; Nelson 1973; 1983; Sullivan 1942: 44, 67, for a selection of accounts, and Mithen 1990: chap. 3 for a collation of ethnographic evidence). Of equal interest is the ability to use the sight of one animal or plant, or a change in its behaviour, to inform about future events, as illustrated most effectively by the so-called 'calendar plants' used by the Groote Eylandt islanders (Levitt 1981). As Blurton-Jones and Konner (1976) demonstrated by their study of the !Kung, the knowledge gained by these methods can be at least as accurate in terms of predicting animal whereabouts as that derived from the behavioural ecology of Western scientists.

We must, however, make a distinction here between this type of ecological knowledge and the use of anthropomorphism. Modern hunter-gatherers make extensive use of the latter, effectively attributing human-like minds to animals, and this can also provide effective predictors of behaviour (e.g. see Gubser 1965; Marks 1976; Silberbauer 1981). More generally they use social knowledge to make sense of the natural world, while also engaging in the converse – natural history knowledge to make sense of the social world – in terms of totemism. Indeed, Kennedy (1992) argues that human beings in general are prone to a compulsive anthropomorphizing, and with modern hunter-gatherers the social and natural worlds appear to have no boundaries. I have argued that this 'cognitive fluidity' has arisen in relatively recent times and is a consequence of the evolution of compositional language (Mithen 1996; 2005). Consequently for pre-modern hominins I would expect a natural history intelligence and a social intelligence that are effectively isolated from each other: the natural and social worlds were two distinct entities.

Evidence from archaeology, developmental psychology and anthropology

The key archaeological evidence to support the proposition of a discrete natural history intelligence (or intuitive biology) evolving relatively early in human evolution is the extraordinary diversity of environments and hence foraging strategies that were adopted by hominins. This is especially the case for those after two million years ago when dispersal – or most likely multiple dispersals – into Eurasia occurred (Straus & Bar-Yosef 2001). This was also a period of substantial environmental fluctuation, not just in terms of the major climatic alterations between glacial and inter-glacial stages, but also at a much finer level of temporal resolution.

The evidence from the earliest sites outside of Africa, such as Dmanisi in Georgia (1.7 mya), Ubediya in Israel (1.5 mya) and Atapuerca in Europe (0.8 mya), indicates that the earliest dispersals occurred with an Oldowan-like technology effectively the same as had been used since 2.5 million years ago in Africa. The earliest dispersals were not, therefore, driven by technological advancement. While both constraints and opportunities would have arisen from the particular nature of Pleistocene environmental history (Mithen & Reed 2002), it seems likely that such dispersals were partly a product of new cognitive capacities within the mind of *Homo ergaster*. These are unlikely to have related to language in the sense that we would ordinarily understand it (Mithen 2005, and see below) and most likely related to enhanced natural history intelligence enabling rapid learning about, and effective decision-making in, new environments.

Those individuals who were more able to make effective decisions about which resources to exploit would have gained reproductive advantage. If we follow the logic of evolutionary psychology, this would have been those individuals who had evolved learning and decision mechanisms for extracting the most relevant information from the natural world and then processing it in the most appropriate manner to make decisions that enhanced survival and reproduction. The character of those decisions would have been dependent upon the particular context of the individuals: in some cases it may have been those which maximized energetic efficiency, in others those which minimized risk of predation. Indeed, those individuals who were able to make (meta-)decisions about their decision-making process would have gained the most reproductive advantage.

Homo ergaster was an ancestor of *Homo sapiens*. Two lines of evidence from the modern day indicate that an evolved natural history intelligence/intuitive biology was inherited by *H. sapiens*, and most likely further evolved within the post-*H. ergaster* lineage of this species. Studies by Keil (1989; 1994), Carey (1985; Carey & Spelke 1994) and Atran (1990; 1994) indicate that children are born with an innate understanding of the differences between living things and inanimate objects. Keil's work has shown that very young children have a propensity to attribute an essence to different types of living things and to recognize that a change in manifest appearance does not necessarily reflect a change in kind.

Complementing this evidence from the study of infants is that of anthroplogy in general and ethnobiology in particular. First we can note the apparent propensity to simply observe, discuss, and classify the natural world found amongst people living 'traditional' lifestyles, as I have reviewed for hunter-gatherers elsewhere (Mithen 1990). This, of course, may not be surprising as their livelihoods are dependent upon such knowledge. But what is even more striking is the ease with which this knowledge is acquired and culturally transmitted – human minds appear to be pre-tuned for acquiring and processing information about animals and plants such that virtually no active teaching is required (Atran 1990; 1994).

Formal training appears to be generally rare among hunter-gatherer societies. When it is reported, it tends to be found within North American groups (e.g. Birket-Smith & De Laguna 1938: 162; Gubser 1965: 109; Ray 1963: 106; Tanner 1979: 44). More often, the reference in ethnographies to formal training is a remark that it is lacking. For instance Goodale (1971: 37) stressed the minimal instruction given to young Tiwi foragers. That which is found relates to the use of specific tools such as fish spears. Similarly, the lack of formal education among the !Kung has been emphasized frequently (e.g. Biesele 1976: 307; Blurton-Jones & Konner 1976: 338-9; Draper 1976: 211-12; Lee 1979: 236; Marshall 1976: 95-6). That which does exist again relates to very specific activities such as stalking and tracking and does not occur until adolescence. Education via informal teaching using story-telling and song appears to be much more common, although it tends not to be directed towards any specific activity. The importance of learning fron one's own experience and experiment is frequently remarked upon (e.g. Goodale 1971: 40; Gubser 1965: 220; Hassrick 1964: 318-19; Lee 1979: 707-8; Tanner 1979: 44). This appears to be closely related to the observation and mimicry of adults (Blurton-Jones & Konner 1976: 339; Hassrick 1964: 318-19).

In addition to the evidence from infant development and ethnographic observations regarding knowledge acquisition, a third source of evidence – and perhaps of most significance – is the universally shared principles of ethnobiological classification: when humans are seen in evolutionary perspective we should not be at all surprised that 'there are unmistakable cross-cultural regularities in the structure of folkbiological classification' (Atran 1994: 17; see also Berlin in this volume). Berlin (1992), more specifically, has argued that all cultures utilize a common five-tiered 'taxonomic' hierarchy for classifying plants and animals: kingdom, life-form, intermediate, generic, and specific. This results in cultural diversity owing to the engagement between such universally held principles of classification, the biodiversity of the natural world, and the historical contexts of human communities.

As noted above, effective classification is one of the keys to rapid and effective learning. Those hominin individuals who had to begin classifying from scratch the animals and plants in a new habitat they entered, either via their own dispersal or owing to environmental change, would have been at a severe disadvantage compared to those who had an existing intuitive scheme with which to classify the natural world. The latter would have been able to adapt more rapidly by making more effective foraging decisions. From this evolutionary perspective, we should not be surprised at Berlin's findings and be sympathetic to his interpretation of these as deriving from universally shared principles of ethnobiological classification.

Non-linguistic understandings of the natural world

The adoption of an evolutionary perspective on the human mind questions the priority that has been given to linguistic approaches in ethnobiology, especially those that involve transcribing spoken words of non-literate cultures into written text for an academic readership. As Roy Ellen describes in his introductory essay to this volume, 'Not only is the written word often inadequate to grasp the precise way in which local peoples perceive their environment, but the vocal and verbal dimension itself is insufficient'.

The majority of palaeoanthropologists agree that language as we understand it is a relatively recent means of communication in human evolution, most likely restricted to *Homo sapiens* and appearing less than 200,000 years ago. By 'language as we understand it' I mean a primarily vocal means of communication that can create a potentially infinite number of utterances from a finite set of words and by using a suite of grammatical rules.

The argument for a recent and single origin of language is based on evidence from the archaeological rather than fossil record. The latter suggests that the vocal and aural apparatus for language was present within pre-modern species, such as *Homo heidelbergensis* and *Homo neanderthalensis* (Arensburg, Schepartz, Tillier, Vandermeersch & Rak 1989; Kay, Cartmill & Balow 1998; MacLarnon & Hewitt 1999; Martínez *et al.* 2004), while the large brains of these species suggest that sufficient neural capacity was present for language. However, the absence of symbolic behaviour and the immense cultural stability of these species suggests that they lacked a linguistic capacity of the kind that would drive creative thought and cultural change (Carruthers 2002; Mithen 1996; 2005).

Proto-language of pre-modern humans

Palaeoanthropologists refer to the communication systems of pre-linguistic hominins as 'proto-language'. Views of proto-language range from utterances composed from a limited number of words and lacking any grammatical structure (Bickerton 1996), via utterances that make use of 'proto-grammatical rules' such as 'agent first' (Jackendoff 1999), to notions of holistic utterances – multi-syllabic phrases with fixed meanings but which were not composed from discrete words (Wray 1998). As we are dealing with at least two million years of human evolution and multiple species and lineages, it is most likely that various forms of proto-language evolved, with that of the earliest hominins being similar to the vocalizations of the African apes today (Mithen 2005).

The critical argument for this paper is that hominins using proto-language were evidently effective at classifying the natural world, acquiring information about animals and plants, and then making behavioural decisions by the use of dedicated mental modules, as described above. Had they not been able to do so, we would not find evidence for the dispersal out of Africa soon after two million years ago and the successful exploitation of diverse environments. The implication is that perceptual, classificatory, and decision-making systems are not dependent upon language; hence by prioritizing linguistic approaches to ethnobiology, we may be restricting our understanding of how humans perceive, understand, and communicate about the natural world.

The Neanderthals are likely to have been non-linguistic hominins that had a profound understanding of the natural world – whether or not one wishes to attribute them with a discrete natural history intelligence (cf. Mithen 1996). They lived in Europe and the Near East between c.250,000 and 28,000 years ago, surviving through the severe climatic fluctuations of the late Pleistocene prior to their extinction. They were big game hunters using stone-tipped spears but their technology appears extremely limited in comparison to that of the *Homo sapiens* that post-dated them in Europe during the so-called 'Upper Palaeolithic' (Mellars 1996; Stringer & Gamble 1983). The relatively unsophisticated character of Neanderthal technology suggests that they were more dependent than are modern humans not only upon physiological adaptations to glacial environments but also upon natural history intelligence to make foraging decisions.

Elsewhere (Mithen 2005) I have followed Wray by proposing that Neanderthal proto-language was holistic and argued that their vocalizations had a high degree of musicality, making extensive use of variations in melody and rhythm to express emotion and induce emotional states in others. As such, Neanderthal vocal and gestural communication should be thought of as proto-music as much as proto-language. While 'messages' would have consisted of arbitrary strings of syllables, Neanderthal utterances probably made extensive use of vocal imitation of natural sounds, especially animal calls, onomatopoeias, and to have involved the phenomenon of sound synaesthesia, as described by Berlin (2005, and in this volume) and which continues to form a key feature of human language. Also, as Donald (1991) has argued, mimesis is likely to have played a key role in hominin communication.

Hmmmmm and the origin of language

I have characterized the type of communication system outlined in the previous section as 'Hmmmmm': *H*olistic, *m*anipulative, *m*ulti-*m*odal, *m*usical, *m*imetic (Mithen 2005). If it was present amongst Neanderthals, then it was also possibly present amongst our immediate ancestors in Africa, a species that some refer to as *Homo helmei*. The process by which this system evolved into 'language as we know it' was described by Wray (1998; 2000) as 'segmentation'. By this she refers to the division of holistic phrases into separate units, each of which had their own referential meaning and could then be re-combined with units from other utterances to create any infinite array of new utterances. This is the emergence of compositionality, the feature that makes language so much more powerful than any other communication system.

Wray suggests that segmentation may have arisen from the recognition of chance associations between the phonetic segments of the holistic utterance and objects or events to which they related (see Bickerton 2003 for a critique). Once recognized, these associations might then have been used in a referential fashion to create the new, linguistic phrases. Support for this idea comes from the use of computer models to simulate the evolution of language (e.g. Batali 2002; Kirby 2000; 2002; Komarova & Nowak 2003; Nowak & Komarova 2001).

The kick-start to segmentation may have been a chance genetic mutation. This may have provided the ability to identify phonetic segments in a holistic utterance, an ability that had previously been absent. Some aspects of language are dependent on the possession of the specific gene FOXP2, the modern human version of which seems to have appeared in Africa at soon after 200,000 years ago (Enard *et al.* 2002). So, at the risk of over-simplifying, it is possible that the process of segmentation was dependent upon this gene in some manner that has yet to be discovered. Indeed, it may be significant that those afflicted by a faulty version of the FOXP2 gene have had difficulties not only with grammar, but also with understanding complex sentences and judging whether a sequence such as 'blonterstaping' is a real word (Bishop 2002). These difficulties seem to reflect a problem with the segmentation of what would have sounded to them to be holistic utterances. So perhaps it was only with the chance mutation of the FOXP2 gene to create the modern human version that segmentation became possible. Alternatively, there may have been other genetic mutations at a similar date that enabled the transition from holistic phrases to compositional language.

Once the process of segmentation had begun, we should expect a rapid evolution of grammatical rules. Such rules would have evolved via the process of cultural transmission in the manner Kirby (2000) describes, and perhaps through natural selection leading to the appearance of genetically based neural networks that enable more complex grammatical constructions. I have argued (Mithen 2005) that the segmentation of Hmmmmm led to the appearance of the two vocal (and part-gestural) communication systems of modern humans: language and music. Their partly shared evolutionary history is evident from the significant overlaps between language and music, while several features of Hmmmmm remain within spoken language. The most notable of these are onomatopoeia, mimicry, and sound synaesthesia, while gesture and body language remain as fundamental aspects of human communication – although we are often entirely unaware of their use and significance (Beattie 2003).

Cognitive fluidity, totemism and anthropomorphism

While the evolution of compositional language via the segmentation of Hmmmmm would have enhanced communication, it may have also led to a more fundamental change in the character of human thought – the transition from a domain-specific to a cognitively fluid mentality.

I have argued elsewhere (Mithen 1996) that cognitive fluidity was a consequence of language: spoken and imaginary utterances acted as conduits for ideas and information to flow from one intelligence to another, a view supported by Carruthers (2002). He drew on the latest research in neuroscience and psychology to argue that the 'imagined sentences' that we create in our minds allow the outputs from one intelligence/module to be combined with those from one or more others, and thereby create new types of conscious thoughts. Carruthers placed considerable emphasis on syntax – an essential part of compositional language. Syntax allows for the multiple embedding of adjectives and phrases, the phenomenon of recursiveness. According to Carruthers, syntax allows one imaginary sentence generated by one type of cognitive module/intelligence to be embedded into that of another imaginary sentence coming from a different module/intelligence. By so doing, a single imaginary sentence will be created which generates an 'inter-modular' or cognitively fluid thought, one that could not have existed without compositional language.

With regard to thought about the natural world, cognitive fluidity allowed this to be integrated with that about the human social world and artefacts. The former allowed the possibility of anthropomorphism – attributing animals with human-like beliefs and desires – and totemism in terms of attributing humans with animal ancestors (Mithen 1996). This type of thinking does not, however, compromise the ability to evaluate the costs, benefits, and risks of different types of actions relating to animals and plants that enable adaptation to the natural world. This is, after all, a cognitive capacity that has a far longer evolutionary history within the human mind than that for anthropomorphic and totemic thinking.

Conclusion

Homo sapiens is the product of six million years of evolution since the time of the common ancestor we shared with the chimpanzee. That evolutionary history has a profound impact on the way we think and behave today, especially with regard to the natural world. Unless we take our evolutionary past into account, we will never get beyond a partial understanding of what it means to be human. I have stressed two impacts of this evolutionary perspective on the study of ethnobiology. First, it is to be expected that whatever their cultural, ecological, or historical context, humans will share a common means for classifying the natural world, this being part of an intuitive biology that is encoded within our genome. Such intuitive biology may have little opportunity to develop for those individuals who grow up with limited contact with the natural world – as with other potential neural capacities, 'use it or lose it' is the message. Nevertheless, the search for underlying shared principles of classification and cognitive engagement with the natural world must be a proper and key feature of ethnobiology.

A second lesson from our evolutionary past is that language is a relatively recent means of human communication, one most likely restricted to our species alone and appearing no longer than two hundred thousand years ago. For almost the whole of the six million years of human evolution since the common ancestor, hominins have perceived, classified, interpreted, and made decisions about the natural world in a pre-linguistic mode. Some of those hominins may have had sophisticated communications system of the type I have characterized as Hmmmmm, which made extensive use of onomatopoeia, mimicry, and sound synaesthesia. Traces of these remain within our language (Berlin 2005, and this volume) and music today (Mithen 2005) but 'language as we know it' is likely to play a relatively minor role in the development and cultural transmission of knowledge about the natural world. So while linguistic approaches to ethnobiology are unquestionably important, these will lead to only a partial understanding of how humans perceive and understand the natural world.

NOTE

I am most grateful to Roy Ellen for inviting me to contribute this paper and for his comments on a draft manuscript.

REFERENCES

AIELLO, L.C. 1996. Terrestriality, bipedalism and the origin of language. In *Evolution of social behaviour patterns in primates and man* (eds) W.G. Runciman, J. Maynard-Smith & R.I.M. Dunbar, 269-90. Oxford: University Press.
——— & R.I.M. DUNBAR 1993. Neocortex size, group size, and the evolution of language. *Current Anthropology* **34**, 184-93.
——— & P. WHEELER 1995. The expensive-tissue hypothesis. *Current Anthropology* **36**, 199-220.
ARENSBURG, B., L.A. SCHEPARTZ, A.M. TILLIER, B. VANDERMEERSCH & Y. RAK 1989. A reappraisal of the anatomical basis for speech in Middle Palaeolithic hominids. *American Journal of Physical Anthropology* **83**, 137-56.

ATRAN, S. 1990. *Cognitive foundations of natural history: towards an anthropology of science.* Cambridge: University Press.
——— 1994. Core domains versus scientific theories: evidence from systematics and Itza-Maya folkbiology. In *Mapping the mind: domain specificity in cognition and culture* (eds) L.A. Hirschfeld & S.A. Gelman, 316-40. Cambridge: University Press.
BATALI, J. 2002. The negotiation and acquisition of recursive grammars as a result of competition among exemplars. In *Linguistic evolution through language acquisition: formal and computational models* (ed.) E. Briscoe, 111-72. Cambridge: University Press.
BEATTIE, G. 2003. *Visible thought: the new language of body language.* London: Routledge.
BERLIN, B. 1992. *The principles of ethnobiological classification.* Princeton: University Press.
——— 2005. 'Just another fish story?' Size-symbolic properties of fish names. In *Animal names* (eds) A. Mineli, G. Ortalli & G. Sanga, 9-21. Venice: Istituto Veneto di Scienze, Lettere ed Arti.
BICKERTON, D. 1996. *Language and human behaviour.* London: UCL Press.
——— 2003. Symbol and structure: a comprehensive framework for language evolution. In *Language evolution* (eds) M.H. Christiansen & S. Kirby, 77-93. Oxford: University Press.
BIESELE, M. 1976. Aspects of !Kung folklore. In *Kalahari hunter-gatherers* (eds) R. Lee & I. DeVore, 302-25. Cambridge, Mass.: Harvard University Press.
BIRKET-SMITH, K. & F. DE LAGUNA 1938. *The Eyak Indians of the Copper River delta, Alaska.* Copenhagen: Levin & Munksgaard.
BISHOP, D.V.M. 2002. Putting language genes in perspective. *Trends in Genetics* 18, 57-9.
BLURTON-JONES, N. & M.J. KONNER 1976. !Kung knowledge of animal behaviour. In *Kalahari hunter-gatherers* (eds) R. Lee & I. DeVore, 326-48. Cambridge, Mass.: Harvard University Press.
BROWN, P., T. SUTIKNA, M.J. MORWOOD, R.P. SOEJONO, JATMIKO, E. WAYHU SAPTOMO & ROKUS AWE DUE 2004. A new small-bodied hominin from the Late Pleistocene of Flores, Indonesia. *Nature* 431, 1055-61.
BYRNE, R.W. 1996a. *The thinking ape.* Oxford: University Press.
——— 1996b. Relating brain size to intelligence in primates. In *Modelling the early human mind* (eds) P. Mellars & K. Gibson, 49-56. Cambridge: McDonald Institute Monographs.
——— & A. WHITEN (eds) 1988. *Machiavellian intelligence: social expertise and the evolution of intellect in monkeys, apes and humans.* Oxford: Clarendon Press.
——— & ——— 1992. Cognitive evolution in primates: evidence from tactical deception. *Man* (N.S.) 27, 609-27.
CAREY, S. 1985. *Conceptual change in childhood.* Cambridge, Mass.: Bradford/MIT Press.
——— & E. SPELKE 1994. Domain-specific knowledge and conceptual change. In *Mapping the mind: domain specificity in cognition and culture* (eds) L.A. Hirschfeld & S.A. Gelman, 169-200. Cambridge: University Press.
CARRUTHERS, P. 2002. The cognitive functions of language. *Behavioral and Brain Sciences* 25, 657-726.
——— & A. CHAMBERLAIN (eds) 2000. *Evolution and the human mind: modularity, language and meta-cognition.* Cambridge: University Press.
——— & P. SMITH (eds) 1996. *Theories of theories of minds.* Cambridge: University Press.
COSMIDES, L. & J. TOOBY 1994. Origins of domain specificity: the evolution of functional organization. In *Mapping the mind: domain specificity in cognition and culture* (eds) L.A. Hirschfeld & S.A. Gelman, 85-116. Cambridge: University Press.
DEMENOCAL, P.B. 1995. Plio-Pleistocene African climate. *Science* 270, 53-9.
D'ERRICO, F. & M.F. SANCHEZ GONI 2003. Neanderthal extinction and the millennial scale climatic variability of OIS 3. *Quaternary Science Reviews* 22, 769-88.
———, J. ZILHÃO, M. JULIAN, D. BAFFIER & J. PELEGRIN 1998. Neanderthal acculturation in Western Europe. *Current Anthropology* 39, S1-S44.

DONALD, M. 1991. *Origins of the modern mind*. Cambridge, Mass.: Harvard University Press.
DRAPER, P. 1976. Social and economic constraints on child life among the !Kung. In *Kalahari hunter-gatherers: studies of the !Kung and their neighbours* (eds) R. Lee & I. DeVore, 199-218. Cambridge, Mass.: Harvard University Press.
DUNBAR, R.I.M. 1993. Coevolution of neocortical size on group size in primates. *Journal of Human Evolution* **20**, 469-93.
——— 1996. *Gossip, grooming and language*. London: Faber & Faber.
——— 1998. Theory of mind and the evolution of language. In *Approaches to the evolution of language* (eds) J.R. Hurford, M. Studdert-Kennedy & C. Knight, 92-110. Cambidge: University Press.
——— 2004. *The human story*. London: Faber & Faber.
ENARD, W., M. PRZEWORSKI, S.E. FISHER, C.S. Lai, V. WIEBE, T. KITANO, A.P. MONACO & S. PAABO 2002. Molecular evolution of FOXP2, a gene involved in speech and language. *Nature* **418**, 869-72.
GABUNIA, L., A. VEKUA, D. LORDKIPANIDZE, C.C. SWISHER, III, R. FERRING, A. JUSTUS, M. NIORADZE, M. TVALCHRELIDZE, S.C. ANTÓN, G. BOSINSKI, O. JÖRIS, M.-A. DE LUMLEY, G. MAJSUARDZE A. MOUSKHELISHVLLI 2000. Earliest Pleistocene hominid cranial remains from Dmanisi, Republic of Georgia: taxonomy, geological setting and age. *Science* **288**, 1019-25.
GAMBLE, C. 1999. *The Palaeolithic societies of Europe*. Cambridge: University Press.
GARDNER, H. 1983. *Frames of mind: the theory of multiple intelligences*. New York: Basic Books.
——— 1993. *Multiple intelligences: the theory in practice*. New York: Basic Books.
GOODALE, J.C. 1971. *Tiwi wives: a study of the women of Melville Island, North Australia*. Seattle: University of Washington Press.
GUBSER, N.J. 1965. *The Nunamiut Eskimos: hunters of caribou*. New Haven: Yale University Press.
HASSRICK, R.B. 1964. *The Sioux*. Norman: University of Oklahoma Press.
HENSHILWOOD, C.S., F. D'ERRICO, M. VANHAEREN, K. VAN NIEKERK & Z. JACOBS 2004. Middle stone age shell beads from South Africa. *Science* **304**, 404.
———, ———, R. YATES, Z. JACOBS, C. TRIBOLO, G.A.T. DULLER, N. MERCIER, J.C. SEALY, H. VALLADAS, I. WATTS & A.G. WINTLE 2002. Emergence of modern human behaviour: Middle Stone age engravings from South Africa. *Science* **295**, 1278-80.
HILL, K. & K. HAWKES 1983. Neotropical hunting among the Áche of Eastern Paraguay. In *Adaptive responses of native Amazonians* (eds) R. Hames & W. Vickers, 139-88. New York: Academic Press.
HIRSCHFELD, L.A. & S.A. GELMAN (eds) 1994. *Mapping the mind: domain specificity in cognition and culture*. Cambridge: University Press.
HUMPHREY, N. 1984. *Consciousness regained*. Oxford: University Press.
INGMAN, M., H. KAESSMANN, S. PAABO & U. GYLLENSTEN 2000. Mitochondrial genome variation and the origin of modern humans. *Nature* **408**, 708-13.
ISAAC, G. 1983. Bones in contention: competing explanations for the juxtaposition of Early Pleistocene artefacts and faunal remains. In *Animals and archaeology: hunters and their prey* (eds) J. Clutton-Brock & C. Grigson, 3-19. Oxford: British Archaeological Reports, International Series 163.
——— 1989. *The archaeology of human origins* (ed. B. Isaac). Cambridge: University Press.
JACKENDOFF, R. 1999. Possible stages in the evolution of the language faculty. *Trends in Cognitive Sciences* **3**, 272-9.
JOBLING, M.A., M.E. HURLES & C. TYLER-SMITH 2004. *Human evolutionary genetics*. New York: Garland Publishing.
JOHANSEN, D. & B. EDGAR 1996. *From Lucy to language*. London: Weidenfeld & Nicolson.
KAY, R.F., M. CARTMILL & M. BALOW 1998. The hypoglossal canal and the origin of human vocal behaviour. *Proceedings of the National Academy of Sciences* **95**, 5417-9.

KEIL, F. 1989. *Concepts, kinds and cognitive development*. Cambridge, Mass.: MIT Press.
——— 1994. The birth and nurturance of concepts by domains: the origins of concepts of living things. In *Mapping the mind: domain specificity in cognition and culture* (eds) L.A. Hirschfeld & S.A. Gelman, 234-54. Cambridge: University Press.
KENNEDY, J.S. 1992. *The new anthropomorphism*. Cambridge: University Press.
KIRBY, S. 2000. Syntax without natural selection: How compositionality emerges from vocabulary in a population of learners. In *The evolutionary emergence of language: social function and the origins of linguistic form* (eds) C. Knight, M. Studdert-Kennedy & J.R. Hurford, 303-23. Cambridge: University Press.
——— 2002. Learning, bottlenecks and the evolution of recursive syntax. In *Linguistic evolution through language acquisition: formal and computational models* (ed.) E. Briscoe, 173-204. Cambridge: University Press.
KNIGHT, C., C. POWERS & I. WATTS 1995. The human symbolic revolution: a Darwinian account. *Cambridge Archaeological Journal* **5**, 75-114.
KOMAROVA, N.L. & M. NOWAK 2003. Language learning and evolution. In *Language evolution* (eds) M.H. Christiansen & S. Kirby, 317-37. Oxford: University Press.
LAHR, M.M. & R. FOLEY 1994. Multiple dispersals and modern human origins. *Evolutionary Anthropology* **3**, 48-60.
LEE, R.B. 1979. *The !Kung San: men, women and work in a foraging society*. Cambridge: University Press.
LEVITT, D. 1981. *Plants and people: aboriginal uses of plants on Groote Eylandt*. Canberra: Australian Institute of Aboriginal Studies.
LEWIN, R. & R. FOLEY 2003. *Principles of human evolution*. London: Blackwell Scientific Inc.
MCBREARTY, S. & A. BROOKS 2000. The revolution that wasn't: a new interpretation of the origin of modern human behavior. *Journal of Human Evolution* **38**, 453-563.
MCDOUGALL, I., F.H. BROWN & J.G. FLEAGLE 2005. Stratigraphic placement and age of modern humans from Kibish, Ethiopia. *Nature* **433**, 733-6.
MACLARNON, A. & G.P. HEWITT 1999. The evolution of human speech: the role of enhanced breathing control. *American Journal of Physical Anthropology* **109**, 341-3.
MARKS, S.A. 1976. *Large mammals and a brave people: subsistence hunters in Zambia*. Seattle: University of Washington Press.
MARSHALL, L. 1976. *The !Kung of Nyae Nyae*. Cambridge, Mass.: Harvard University Press.
MARTÍNEZ, I., M. ROSA, J.-L. ARSUAGA, P. JARABO, R. QUAM, C. LORENZO, A. GRACIA, J.-M. CARRETERO, J.-M. BERMÚDEZ DE CASTRO & E. CARBONELL 2004. Auditory capacities in Middle Pleistocene humans from the Sierra de Atapuerca in Spain. *Proceedings of the National Academy of Sciences* **101**, 9976-81.
MELLARS, P. 1996. *The Neanderthal legacy*. Princeton: University Press.
——— 2004. Neanderthals and the modern human colonization of Europe. *Nature* **432**, 461-5.
MITHEN, S.J. 1990. *Thoughtful foragers: a study of prehistoric decision making*. Cambridge: University Press.
——— 1996. *The prehistory of the mind: a search for the origin of art, science and religion*. London: Thames & Hudson.
——— 1999. Palaeoanthropological perspectives on the theory of mind. In *Understanding other minds: perspectives from autism and cognitive neuroscience* (eds) S. Baron-Cohen, H.T. Flusberg & D. Cohen, 494-508. Oxford: University Press.
——— 2003. *After the Ice: a global human history, 20,000-15,000 BC*. London: Weidenfeld & Nicolson.
——— 2005. *The singing Neanderthals: the origin of music, language, mind and body*. London: Weidenfeld & Nicolson.

——— & M. REED, M. 2002. Stepping out: a computer simulation of hominid dispersal from Africa. *Journal of Human Evolution* **43**, 433-62.

MORWOOD, M.J., P.B. O'SULLIVAN, F. AZIZ & A. RAZA 1998. Fission-track ages of stone tools and fossils on the east Indonesian island of Flores. *Nature* **392**, 173-6.

———, R.P. SOEJONO, R.G. ROBERTS, T. SUTIKANA, C.S.M. TURNEY, K.E. WESTAWAY, W.J. RINK, J.-X. ZHAO, G.D. VAN DEN BERGH, ROKUS AWE DUE, D.R. HOBBS, M.W. MOORE, M.I. BIRD & L.K. FIFIELD 2004. Archaeology and the age of a new hominin from Flores in eastern Indonesia. *Nature* **431**, 1087-91.

MULVANEY, J. & J. KAMMINGA 1999. *Prehistory of Australia*. Washington, D.C.: Smithsonian Institution Press.

NELSON, R.K. 1973. *Hunters of the Northern forest: designs for survival among the Alaskan Kutchin*. Chicago: University Press.

——— 1983. *Make prayers to the raven: a Koyukon view of the Northern Forest*. Chicago: University Press.

NOWAK, M.A. & N.L. KOMAROVA 2001. Evolution of universal grammar. *Science* **291**, 114-8.

O'CONNELL, J. & J. ALLEN 2004. Dating the colonization of Sahul: a review of recent research. *Journal of Archaeological Science* **31**, 835-53.

PINKER, S. 1997. *How the mind works*. New York: Norton.

POTTS, R. 1988. *Early hominid activities at Olduvai Gorge*. New York: Aldine de Gruyter.

POVINELLI, D.J. 1993. Reconstructing the evolution of the mind. *American Psychologist* **48**, 493-509.

——— 1999. *Folk physics for apes*. Oxford: University Press.

RAY, V.F. 1963. *Primitive pragmatists: the Modoc Indians of northern California*. Seattle: University of Washington Press.

RUFF, C.B., E. TRINKAUS & T.W. HOLLIDAY 1997. Body mass and encephalization in Pleistocene *Homo*. *Nature* **387**, 173-6.

SAMUELS, R. 1998. Evolutionary psychology and the massive modularity hypothesis. *British Journal for the Philosophy of Science* **49**, 575-602.

SCHICK, K. & N. TOTH 1993. *Making silent stones speak: human evolution and the dawn of technology*. New York: Simon & Schuster.

SILBERBAUER, G. 1981. *Hunter and habitat in the central Kalahari Desert*. New York: Cambridge University Press.

SMITH, B.D. 1995. *The emergence of agriculture*. New York: Scientific American Library.

STEELE, J. 1996. On predicting hominid group size. In *The archaeology of human ancestry* (eds) J. Steele & S. Shennan, 230-52. London: Routledge.

STRAUS, L.G. & C. BAR-YOSEF (eds) 2001. Out of Africa in the Pleistocene. *Quaternary International* **75**.

STRINGER, C.B., R.N.E. BARTON & C. FINLAYSON 2000. *Neanderthals at the edge*. Oxford: Oxbow Books.

——— & C. GAMBLE 1983. *In search of the Neanderthals*. London: Thames & Hudson.

SULLIVAN, R.J. 1942. *The Ten'a food quest*. Washington, D.C.: The Catholic University of America Press.

TANNER, A. 1979. *Bringing home animals: religious ideology and mode of production of the Mistassini Cree hunters*. London: C. Hurst.

THIEME, H. 1997. Lower Palaeolithic hunting spears from Germany. *Nature* **385**, 807-10.

TOMASELLO, M., J. CALL & B. HARE 2003. Chimpanzees understand psychological states – the question is which ones and to what extent. *Trends in Cognitive Sciences* **7**, 153-6.

WHEELER, P. 1991. The influence of bipedalism on the energy and water budgets of early hominids. *Journal of Human Evolution* **21**, 107-36.

WHITE, T.D., B. ASFAW, D. DEGUSTA, H. GILBERT, G.D. RICHARDS, G. SUWA & F. CLARK HOWELL 2003. Pleistocene *Homo sapiens* from Middle Awash, Ethiopia. *Nature* **423**, 742-7.
WHITEN, A., J. GOODALL, W.C. MCGREW, T. NISHIDA, V. REYNOLDS, Y. SUGIYAMA, G.E.G. TUTIN, R.W. WRANGHAM & C. BOESCH 1999. Culture in chimpanzees. *Nature* **399**, 682-5.
WOOD, B. 1992. Origin and evolution of the genus *Homo*. Nature **355**, 783-90.
—— & M. COLLARD 1999. The human genus. *Science* **284**, 65-71.
WRAY, A. 1998. Protolanguage as a holistic system for social interaction. *Language and Communication* **18**, 47-67.
—— 2000. Holistic utterances in protolanguage: the link from primates to humans. In *The evolutionary emergence of language: social function and the origins of linguistic form* (eds) C. Knight, M. Studdert-Kennedy & J.R. Hurford, 285-302. Cambridge: University Press.

4

The interplay of ethnographic and archaeological knowledge in the study of past human subsistence in the tropics

DAVID R. HARRIS

Archaeologists concerned with the prehistory of human subsistence, and ethnobiologists who study traditional systems of environmental knowledge in non-industrial societies, have much to learn from each other. But they tend to pursue separate research agendas, and by so doing fail to gain insights that closer collaboration could yield. Their primary sources of data – fragmentary material remains and information from living people – are very different, but potentially complementary. The challenge is to bring together the evidence each discipline generates in a reciprocal relationship that enriches our understanding of how people have interacted with plants and animals.

Attempts to correlate archaeological and ethnobiological data are founded on the assumption that the present is the key to the past, or, in other words, that ethnographically observed practices are a valid guide to interpretation of the bioarchaeological record. But in trying to track archaeologically the antecedents of subsistence practices known from the ethnographically described present or recent past, we may disregard alternative interpretations of enigmatic archaeological evidence. In general, the further back in time we go, the greater the likelihood of a lack of continuity between present and past human subsistence behaviour.

This problem is partially compensated by inferences we can draw from the anatomical, physiological, and behavioural characteristics of the plants and animals, including humans, identified in the archaeological record. For example, it is reasonable to infer, from the evidence of diagnostic butchery marks on the bones of large mammals killed by people at the Lower Palaeolithic site of Boxgrove in Sussex half a million years ago, that bone marrow was a prized human food at that time (Parfitt & Roberts 1999). Similarly, data from ecological studies of present-day mollusc populations on the northwest coast of Sicily, combined with analyses of shells from Mesolithic and early Neolithic occupation layers in nearby caves, indicate that shellfish were an important seasonal food for the inhabitants of the caves 10,000-5,000 years ago (Mannino & Thomas 2003/4; Thomas & Mannino 2003).

Nevertheless, the shorter the time gap between ethnographically observed and archaeologically inferred subsistence practices, the more confident we can be in interpreting the history of those practices. It is not surprising that the subdiscipline of ethnoarchaeology, which seeks to combine observation of present-day human activities with the recovery of archaeological evidence of such activities, has been most successful when the time gap is relatively short. Research on agricultural history in highland Papua New Guinea provides an example of how rewarding this approach can be. When Jack Golson began his archaeological investigations of early agriculture in the swamplands of the Waghi valley in the 1960s, he discovered well-preserved remains of wooden digging sticks. The implements were recognized by his Papuan assistants as the type of paddle-shape digging tools made by their ancestors to ditch and mound the swampy soils for root-crop cultivation, before steel spades were introduced in the twentieth century. By radiocarbon dating the wooden tools, Golson was able to demonstrate, for the first time, that agriculture in highland New Guinea had an antiquity of at least 2,300 years (Golson, Lampert, Wheeler & Ambrose 1967; Lampert 1967). Now, after more than four decades of ethnoarchaeological research in the Waghi valley by Golson and his colleagues, we know that crop cultivation in highland New Guinea dates back some 10,000 years (Denham, Haberle & Lentfer 2004; Denham *et al.* 2003; Golson 1997).

This example shows how successful a project can be that incorporates ethnographic and archaeological evidence in a reciprocal relationship to build and test hypotheses about the past. Unfortunately situations where this is possible have rapidly diminished in recent decades as modernization has quickened and penetrated ever more deeply into less developed areas of the world. Historical evidence of past subsistence practices can sometimes amplify or take the place of ethnographic evidence. It is particularly valuable in regions, such as parts of North America and Australia, where ethnographic data were recorded, systematically or more casually, before traditional subsistence practices were replaced under the impact of modernization (see, e.g., Allen 1974; Cane 1989; Hallam 1989; Harris 1984; R. Jones & Meehan 1989; Lawton, Wilke, DeDecker & Mason 1976; O'Connell, Latz & Barnett 1983; Steward 1930). But valuable as such historical evidence often is, it tends to be frustratingly incomplete, and those who use it to interpret past practices need to be aware of, and make allowance for, the cultural mindsets of the original observers.

These benefits and limitations of the ethnographic and historical evidence are paralleled by the merits and deficiencies of the archaeological evidence. It too is often frustratingly inadequate, for example in providing direct evidence of methods of cultivation and of how foods were processed and consumed. Also, poorly preserved organic remains often cannot be identified by their external morphology, either visually or using conventional microscopy. However, in recent decades, several novel techniques have become available that use microanatomical, histological, and cytological criteria to identify fragmentary and amorphous plant remains – and they are increasingly being applied to the analysis of such remains from archaeological sites, particularly in the tropics (Hather 1994). This trend, coupled with the ability of AMS (accelerator mass spectrometric) radiocarbon dating to determine the age of individual seeds and other very small samples of organic material (Gowlett & Hedges 1986; Harris 1987a), is adding even greater interdisciplinary breadth to the already wide spectrum of ethnobiology. It has the capacity to transform research on past human subsistence, across the entire continuum of interaction between people, plants, and animals that stretches from the procurement of wild resources by hunter-fisher-gatherer foragers through the exploitation of selected plants and animals in systems of 'low-level food production' (Smith 2001) to agriculture based on domesticated crops and livestock raised by farmers and pastoralists (Harris 1996: 444-56).

In addition to these new approaches, conventional ethnoarchaeological methods remain valid means of investigating past subsistence practices, for example comparing field evidence of subsistence-related structures, such as field boundaries, terraces, irrigation and drainage channels, fish traps, and food-processing and -storage sites, with information derived from ethnographic and historical accounts. In this contribution, my purpose is to illustrate some of these novel and more conventional approaches by summarizing aspects of research on two themes: the antiquity of root and tuber cultivation in parts of the American, African, and Southeast Asian tropics; and past forager subsistence in tropical northeastern Australia. In the first part I discuss mainly research undertaken recently by other workers, whereas the second is based on field and ethnohistorical research I undertook in the 1970s and 1980s.

Tropical root and tuber cultivation
It has long been assumed by students of early agriculture in the tropics that vegetatively reproduced root and tuber crops, such as yams, taro, sweet potato, and manioc, have a very long history of cultivation and were probably among

the earliest tropical food plants to be cultivated and domesticated (Harris 1972; 1973; Lathrap 1977; Sauer 1936; 1952: 45-8). This view rested on ethnographic and historical evidence of their place in the tropical agricultural systems of the American, African, South and Southeast Asian tropics, which showed that they were commonly more important staples than seed-reproduced cereals, pulses, or tree crops. The supposition that root and tuber cultivation was a very ancient practice was also based in part on the dubious assumption that, because vegetative propagation was thought to be simpler than raising plants from seed, it was likely to have preceded seed-crop cultivation.

Archaeological evidence was needed to substantiate these ethnohistorically derived inferences, but the soft tissues of roots and tubers decay quickly and their remains seldom survive in a condition that allows them to be identified from gross morphological features. Whole root and tuber organs or fragments thereof have occasionally been found in very dry, cold, or waterlogged contexts sufficiently well preserved to be identified (see, e.g., Rosendahl & Yen 1971; Ugent, Pozorksi & Pozorski 1982; 1986), but this happens too rarely to contribute significantly to our knowledge of the history of agriculture in the tropics. The hard-coated seeds of cereals, pulses, and other seed crops survive much better than roots and tubers and are more commonly recovered, by dry sieving or flotation, in the course of excavation. This bias in the archaeological record in favour of such seed crops as maize, rice, sorghum, and beans has made it particularly difficult to test hypotheses about the importance of roots and tubers in early tropical agriculture. But this unsatisfactory situation is now changing, as the development and application of novel techniques of identification begin to generate direct evidence of the ancient cultivation of root and tuber crops. In particular, three techniques – parenchyma, phytolith, and starch-grain analysis – have the capacity to add substantially to our present meagre knowledge of the prehistory of root-crop cultivation.

Parenchyma analysis
Parenchyma analysis was pioneered at the Institute of Archaeology in London by Jon Hather (1988; 1991; 1994), who succeeded in demonstrating that vegetative parenchymous storage organs (which store carbohydrates and sugars in swollen roots and stems) could survive charring and that the tissues and individual cells were sufficiently diagnostic to allow several root and tuber taxa to be identified to genus or even species level. By using scanning electronic microscopy to compare experimentally charred samples of known taxa with charred fragments from various tropical archaeological sites, he was able to identify the remains of a species of yam (*Dioscorea bulbifera*) from Samoa (Hather 1994: 55-6), sweet potato from Mangaia Island in central Polynesia (Hather & Kirch 1991), and manioc from the Mayan site of Cuello in Belize

(Hather & Hammond 1994). These results demonstrated the feasibility of the technique, and they produced direct evidence of the pre-Columbian presence of the sweet potato (of South American origin) in central Polynesia and of the cultivation of manioc in the Pre-classic period in Central America. Following the success of these early results, Hather trained several archaeobotanists in parenchyma analysis and published a guide to the technique (Hather 2000). It is beginning to be applied more widely, for example by Fuller, Korisettar, Venkatasubbaiah, and Jones (2004: 125-6) in southern India, but the great contribution it could make to our knowledge of the history of root and tuber cultivation in the tropics as a whole is far from being realized.

Phytolith analysis
Phytolith analysis has also begun to produce direct evidence of early root and tuber cultivation in the tropics. Phytoliths are silicified particles of plant tissue that retain the shape of individual cells and can often be identified to the level of family, genus, or, sometimes, species. They resist decay and are widely preserved in soils and sediments in temperate as well as tropical environments. The use of phytolith analysis in archaeology was pioneered in the 1980s, particularly by the American archaeobotanists Dolores Piperno (1988) and Deborah Pearsall (1994), and it is now a well-established technique used in archaeobotanical and geoarchaeological research. It is especially valuable in investigations of early agriculture in the humid tropics, where macroscopic remains of plants tend to survive less successfully than they do in drier tropical and temperate environments.

Initially, phytolith specialists focused their investigations of early agriculture on seed crops, particularly maize (see, e.g., Piperno 1989: 545-7; Piperno, Husum Clary, Cooke, Ranere & Weiland 1985), but more recently Piperno has identified phytoliths of two tropical American tuber crops – arrowroot (*Maranta arundinacea*) and leren (*Calathea allouia*) – in archaeological deposits at a site (the Aguadulce rockshelter) in central Panama (Piperno & Pearsall 1998: 213-17). Both plants belong to the indigenous, ethnographically described root-crop complex of the tropical American lowlands, the antiquity of which has, in the absence of direct evidence, been the subject of much speculation. These new findings indicate that at least those two tuber species were being cultivated in Central America 7,000 years ago, and demonstrate the potential value of phytolith analysis in research on the beginnings of root-crop cultivation in the American tropics.

Phytolith analysis has yet to make a significant contribution to the investigation of early root-crop cultivation in the African or Southeast Asian tropics, but the technique has yielded new evidence of the antiquity of banana cultivation in both regions. Ethnographic and historical data show that bananas of the genus *Musa* were an integral part of the associations of vegetatively reproduced plants (most of which are root and tuber crops) cultivated in historical times in the humid tropics of Southeast Asia, Africa, and South America, but the ancestral wild forms occur only in southeastern Asia from eastern India to New Guinea and in northeastern Australia (Simmonds 1976). Phytoliths of banana (*Musa* sp., of the Eumusa and Ingentimusa sections of the genus) were first identified at the Kuk archaeological site in the Waghi valley in highland Papua New Guinea by Wilson (1985); and Bowdery (1999) detected at Kuk a change from phytoliths characteristic of wet-habitat plants to those typical of drier habitats that may relate to the history of drainage of the swamp for banana and root-crop cultivation (Denham, Haberle & Lentfer 2004; Denham *et al.* 2003). These data suggest that bananas were being cultivated at Kuk by about 7,000 years ago, and that they were locally domesticated rather than, as had previously been assumed, introduced to New Guinea from mainland Southeast Asia. This conclusion is independently supported by the results of genetic research (Lebot 1999: 621-2), and adds an interesting new dimension to the long-debated but unresolved question of when and by what routes *Musa* bananas were introduced to tropical Africa and South America.

Some light has recently been shed on this question by the recovery of phytoliths of banana type at the archaeological site of Nkang in Cameroon dated to about 2,500 years ago (Mbida *et al.* 2001; Mbida, van Neer & Vrydaghs 2000). The phytoliths are identified as of *Musa* type, rather than from the allied banana genus *Ensete*, which is native to East Africa. The report of phytoliths from Nkang has been questioned (Vansina 2004), and re-affirmed (Mbida *et al.* 2004), and their discovery appears to constitute the first direct archaeobotanical evidence of banana cultivation in tropical Africa. Phytolith analysis has not yet yielded any comparable evidence of banana cultivation in South America, but it has the potential to do so and thus help to resolve the question of whether bananas were introduced before Europeans began to colonize the continent after AD 1500 – an untested inference based on the importance of bananas in traditional systems of swidden and house-garden cultivation in Amazonia – or, as is more usually assumed, reached the area after AD 1500 and were rapidly adopted because they were a valuable resource well suited ecologically to integration into the pre-existing suite of root and tree crops (see, e.g., Harris 1971).

Starch-grain analysis
Starch-grain analysis is the third novel technique that promises to contribute to our knowledge of early root and tuber cultivation in the tropics. It depends on the fact that starch grains occur in plants in many forms, that they tend to survive in a wide variety of depositional environments, and that they are

diagnostic of genus and sometimes species. They are of particular interest to archaeologists because they are often preserved in organic residues found adhering to stone tools (Loy 1994). Their systematic study in archaeological contexts has only recently begun, but the potential of the technique has already been demonstrated by Piperno and others. In Panama she and her colleagues (Piperno & Holst 1998; Piperno, Ranere, Holst & Hansell 2000) have identified arrowroot starch grains embedded in grinding stones excavated at the Aguadulce rockshelter, where the phytoliths of arrowroot and leren were recovered, and also starch grains of manioc and yams (probably wild yam species and also the domesticated American yam *Dioscorea trifida*), all from preceramic stratigraphic horizons dated to between 7,000 and 6,000 years ago.

Starch residues on stone tools from a site in southern Venezuela have also recently been investigated by Linda Perry (2002; 2004). She has analysed starch grains preserved on microlithic flakes that functioned as the 'teeth' of ceramic grater boards found at the site of Pozo Azul Norte-1 in the middle Orinoco basin, dated to c.AD 740. She was able to identify securely starch grains from three plants with edible underground storage organs: yam, arrowroot, and guapo (*Myrosma* sp.), and also maize, the grains of which exceeded those from any other taxon. Surprisingly, no grains of manioc starch were found – a result that undermines the widespread assumption in the archaeological literature on prehistoric agriculture in Amazonia that ceramic graters and associated microliths are reliable indicators of bitter-manioc cultivation. Here we have an interesting example of how a new archaeobotanical technique can produce data that fail to support a long-held inference derived originally from ethnographic observations and historical accounts. These results do not falsify the assumption that ceramic graters armed with microliths were used in prehistoric times to process bitter manioc, but they do reveal that these artefacts have been used to process a wider variety of starch-yielding plants, including maize, and that archaeological evidence of them should no longer be uncritically regarded as a proxy indicator of manioc cultivation.

These data from analyses of starch grains and phytoliths in Panama and Venezuela demonstrate the potential value of both techniques for overcoming the general lack of direct archaeobotanical evidence, and they currently provide the earliest evidence of root-crop cultivation in the Americas. Starch-grain analysis has not yet yielded any definite evidence of early root-crop cultivation in the Southeast Asian or African tropics, but at the site of Niah Cave in Sarawak (Barker 2002) recent analyses by Huw Barton and Victor Paz (in press; Barton 2005) of starch grains from tubers and rhizomes, and charred parenchyma tissues from tubers (and fruits), have produced evidence of the exploitation of tuberous plants such as taros and yams in the Late Pleistocene. Two of the parenchyma fragments have been identified as from the highly toxic yam *Dioscorea hispida*, the large tubers of which have to be carefully processed before they can safely be consumed (Burkill 1966 [1935]: 832). These results demonstrate very early, pre-agricultural use of tubers by rain-forest foragers, presumably for food and perhaps also, in the case of *D. hispida*, for poisons used in hunting, a use of this species recorded ethnographically elsewhere in Southeast Asia by Burkill (1966 [1935]: 831-4).

As all these new techniques come to be more widely applied, we can expect to learn much more about the prehistory of root and tuber cultivation in the American, African, South and Southeast Asian tropics and its relationship to the early cultivation of such seed crops as maize, rice, sorghum, and other millets. But it is more difficult to determine whether starch grains, phytoliths, and fragments of parenchyma derive from wild plants or (morphogenetically) domesticated crops than it is when identifying seeds and other macroscopic plant remains. Nevertheless, by comparing parenchyma, phytoliths, and starch grains of wild and domestic living plants, it is possible to establish criteria, such as larger cell and grain size, by which to identify the domestic forms. All three techniques also have the potential to generate new evidence of past tropical forager subsistence, as is demonstrated by the discovery and identification of parenchyma fragments and starch grains of Late Pleistocene age at Niah Cave.

Tropical forager subsistence

The novel techniques discussed in the previous section can be expected to make important contributions in the future to investigations of plant use by past tropical hunter-fisher-gatherers, but, as was pointed out in the introduction, much can still be learned through the use of more conventional ethnoarchaeological methods based on the comparison of field evidence with ethnographic and historical data. In the second half of this paper, I illustrate this point by summarizing research that I have undertaken on pre-European subsistence in two contrasted areas in tropical northeastern Australia: the north-Queensland rain forest and the western Torres Strait Islands.

Pre-European subsistence in the northeast Australian rain forest

The aim of my research in the Queensland rain forest (Harris 1978; 1987b) was to examine the relationship between historical accounts of Aboriginal life in the forest, archaeological evidence, and rain-forest ecology. When Europeans first penetrated the forest in the late nineteenth century they encountered tribal groups who differed physically and culturally from the Aboriginal inhabitants of the adjacent areas of savanna woodland, and who were living at higher population densities entirely within the forest (i.e. at $c.2\,\text{km}^2$/person). It is clear from the early European descriptions, and confirmed by more recent testimony of descendants of the Aboriginal forest dwellers, that tree nuts were a staple food, systematically harvested, stored, and processed. Fish from the many rivers that traverse the forest were also an important source of food, with terrestrial animals contributing less to Aboriginal diet.

By combining fieldwork, study of the Aboriginal plant names recorded by early European observers, and correlation of the names with the scientific binomials and descriptions provided by the botanist who published a catalogue and flora of Queensland plants at the beginning of the twentieth century (Bailey 1899-1905; 1909), I was able to identify most of the plants reported to have been exploited for food in the rain forest. The historical accounts contain many references to the dietary importance of tree nuts, and I succeeded in identifying eleven tree species that were staple or supplementary sources of edible nuts (Table 1). However, the kernels of seven of these species, six of which appear to have been staple foods, are sufficiently toxic to require laborious processing, by pounding, grating, and leaching in water, before they can safely be consumed. I therefore turned to the question of why they made such an important contribution to Aboriginal subsistence.

Table 1. Principal nut-yielding trees that contributed to Aboriginal diet in the northeast Australian rain forest.

Family and binomial	English name	Inferred status	Leached
Lauraceae			
Beilschmiedia bancroftii	Yellow walnut	PS	X
Endiandra palmerstonii	Black walnut	PS	X
Endiandra pubens	Hairy walnut	SS	X
Endiandra tooram	Brown walnut	SS	X
Elaeocarparceae			
Elaeocarpus bancroftii	Queensland almond	PS	
Proteaceae			
Macadamia whelanii	Silky oak	SS	X
Hicksbeachia pinnatifolia	Ivory silky oak	SS	X
Helicia diversifolia	White oak	SS	
Euphorbiaceae			
Aleurites moluccana	Candlenut	SP	
Leguminosae			
Castanospermum australe	Black bean	SP	X
Podocarpaceae			
Podocarpus amarus	Black pine	SP	

PS = primary staple; SS = secondary staple; SP = supplementary source of starchy food (modified from Harris 1987b: Table 14.1).

Their importance was revealed not only by the historical accounts but also by the existence of two highly distinctive types of stone tool, both of which were evidently used for nut-processing. One type consisted of a large anvil stone pitted with small spherical hollows, accompanied by a hammer stone, and the other was an ovate or rectangular grooved slab associated with a small crushing stone (Figs 1 and 2). Many of these tools have been found in the forest, and when land has been cleared. The question then arose of which nuts were processed with these tools. The answer was provided partly by consideration of the properties of the nuts themselves, partly by references to the tools in the historical accounts, and partly by the testimony of Aboriginal people who retained knowledge of their former use.

Figure 1. Nut-cracking anvil and hammer stones in the rain forest on Mt Whitfield, near Cairns, north Queensland, used principally to crack the thick-shelled nuts of *Elaeocarpus bancroftii*, two of which have been placed adjacent to the hammer stone with, for comparison, one thin-shelled *Beilschmiedia bancroftii* nut below; note that the anvil stone has hollows on two of its surfaces, indicating that the stone was sometimes turned to allow both to be used for nut-cracking (photo D.R. Harris, September 1974).

Figure 2. Grooved stone slab (*morah*) and hand-held crushing stone (*moogi*) used principally to macerate the toxic kernels of *Endiandra palmerstonii*; the moogi has been turned upside down to show the worn surface caused by rocking it back and forth as the kernels are mashed (Queensland State Forestry Office, Atherton, photo D.R. Harris, September 1974).

Measurement of the average weights of samples of several of the nuts that were staple foods revealed an extreme contrast in the ratio of shell to kernel between the nuts of the yellow walnut (*Beilschmiedia bancroftii*) and the Queensland almond (*Elaeocarpus bancroftii*): respectively 2.9:1.00 and 13.6:1.00. They also differ in another significant way – the yellow walnut kernels are toxic and need to be leached whereas those of the Queensland almond can be eaten raw. These contrasts, coupled with the historical and oral information about nut-processing, indicate the specialized functions of the stone tools: the anvil and hammer stones were used to crack the exceptionally thick shells of the Queensland almond; and the grooved slabs and crushing stones were used to macerate the toxic kernels of the yellow walnut (also those of the black walnut, *Endiandra palmerstonii*) before the resultant mash was leached in water.

The two walnuts have thin, easily removed, shells and were evidently particularly important staple foods in the wet season (Meston 1904: 6). In contrast, the Queensland almond has a very hard, thick shell and a single small kernel, and cracking open the shell to extract the kernel is a laborious task. It therefore seems surprising that the almond too was an important staple. However, closer consideration of its properties suggests why it was a preferred species: first, each fruiting tree yields large numbers of nuts that accumulate on the forest floor and are easily collected; second, their thick shells confer some protection against predation by rodents and ensure that the nuts store well; and third – as nutritional analysis showed – their kernels are rich in vegetable oil, unlike those of the yellow walnut (45.11 vs 0.59%), and they also contain appreciable amounts of carbohydrate and protein. The starch-rich walnuts and oil-rich almonds thus complemented each other nutritionally and offered contrasting tastes in what was a predominantly vegetarian diet. They also provided significant quantities of protein which supplemented the rather meagre supplies of animal protein obtained from fish, birds, reptiles, insects, and small mammals.

This study illustrates how consideration of ecological and (in this case very limited) archaeological data can substantiate and elaborate an interpretation of past forager subsistence that is based mainly on historical evidence. So too does my second example, based on fieldwork in the western Torres Strait Islands.

Pre-European subsistence in the western Torres Strait Islands
In Torres Strait I undertook a comparative study of pre-European subsistence along the chain of islands between Cape York and the Papuan coast (Harris 1977: 439-48; 1979), where historical and ethnographic sources indicated that the importance of crop cultivation as a component of the subsistence economy increased from south to north across the Strait. The inhabitants of the southwestern islands obtained almost all their food from wild animals and plants, principally fish, marine turtles, dugongs, shellfish, wild yams, and mangrove shoots, whereas in the midwestern and northwestern islands small-scale cultivation of yams, taro, sweet potato, sugar cane, bananas, and coconuts was an important subsistence activity.

Fieldwork in the islands in 1974 and subsequent years added ethnobiological and archaeological dimensions to the picture of mid-nineteenth-century economy and society in the western islands that could be drawn from the historical and ethnographic sources. In general, this new information reinforced and added to the ethnohistorical evidence. Thus, we found evidence of former agricultural mounds, ditches, and terraces on the midwestern and northwestern islands but not in the southwestern group, and coastal middens containing fish, turtle, and dugong bones, and the remains of fish traps on offshore reefs, were most abundant in the southwestern and midwestern islands. Our attempts to excavate and date some of the agricultural mounds were only partially successful, largely due to the lack of stratified organic remains that could be radiocarbon dated, but we did obtain a date of *c.*700 years ago for the construction of a mound on the northwestern island of Saibai (Barham & Harris 1985: 261-7).

One particularly interesting example of the interplay of ecological and historical data arose from my ethnobiological fieldwork. There is an exceptionally detailed and revealing account of life on the largest of the southwestern islands, Muralug (Prince of Wales Island), left by a young Scottish woman, Barbara Thompson, who lived with the islanders for four years in the 1840s after being shipwrecked (Moore 1978). She was rescued in October 1849 by the crew of HMS *Rattlesnake* and described her life on Muralug to the expedition artist, Oswald Brierly. She told him how each year during the wet season, when wild yams and most other plant foods were unavailable and it was difficult to procure fish, turtle, or dugong, the whole population of the island congregated on the sheltered south coast close to the only place on the island where a species of mangrove grew that was evidently the main source of food during the wet season (Brierly in Moore 1978: 275). She recalled how unpleasant it was collecting the mangrove 'pods' (actually germinating embryos) and struggling back in the rain with heavy baskets full of them to the long communal hut in which all the families lived (Brierly in Moore 1978: 276).

We tend not to think of mangroves as a source of human food, and may be inclined to doubt the truth of Barbara Thompson's statement that they were the wet-season staple, but during fieldwork at Lockhart on the east coast of the Cape York Peninsula, and also in Papuan coastal villages north of Torres Strait, I was told by informants that the 'fruits' of one species of mangrove (*Bruguiera gymnorhiza*) had been a staple food in the recent past (Harris 1977: 432, 449). This is almost certainly the species that was exploited for food on the island of Muralug. Barbara Thompson described how the embryos had to be processed by steaming them in an earth oven, scraping out the soft tissues, and mashing and leaching them in water to produce an edible pulp that was then moulded by hand into balls before being eaten. Macgillivray (1852), who was the naturalist on board HMS *Rattlesnake*, described the pulp as a grey slimy paste, and later European authors regarded *Bruguiera gymnorhiza* as a famine or lean-period food (Flecker, Stephens & Stephens 1948: 12; Hale & Tindale 1933: 115). However, nutrient analysis of the embryos showed them to contain 4.45 per cent crude protein, which would have made a valuable contribution to the wet-season diet of the islanders.

Archaeological reconnaissance and trial excavations carried out by Moore (1978: 14) in the area of the wet-season settlement on Muralug produced evidence of earth ovens, cooking stones, bone fragments, and the remains of edible shellfish, and a charcoal sample was radiocarbon dated to *c*.600 years ago, but although mangrove embryos had probably been processed in the earth ovens, their soft tissues could not be expected to have survived in recognizable form in the archaeological deposits. It is possible that residues from mangrove-processing could now be identified by phytolith or starch-grain analysis, but the fact that no stone or shell artefacts were found in the course of Moore's excavations suggests that it would be very difficult to demonstrate mangrove-processing unequivocally. Nevertheless, this case study, like that of forager subsistence in the Queensland rain forest, demonstrates how, by evaluating ethnohistorical evidence against data derived from ethnobiological fieldwork, new knowledge can be gained, and, conversely, how the former can inform and guide the latter.

Conclusion

In this chapter I have shown how several novel analytical techniques are being used to identify previously unidentifiable traces of plant remains and are providing valuable new evidence of, especially, early root and tuber cultivation in the tropics. These techniques not only generate wholly new data; they can also be used to evaluate existing inferences derived from ethnobotanical and archaeological evidence, as is shown by Linda Perry's investigation (2002; 2004) of starch grains preserved on microliths from ceramic graters that were previously assumed to be diagnostic of bitter-manioc cultivation but are now known to have been used to process maize as well as manioc and other root crops.

This example is particularly illuminating because it demonstrates the value of using a new archaeobotanical technique to test and refine a pre-existing body of archaeological and ethnographic evidence. It goes beyond the discovery and identification of new assemblages of plant remains (the valuable but limited achievement of much recent research in parenchyma, phytolith, and starch-grain analysis) to reveal new insights into how the plants were used. In recent years, archaeobotanists have increasingly tried to elicit direct evidence of how plants were cultivated and processed, notably through ecological and ethnographic studies of cereals and their associated weed floras, an approach pioneered by Gordon Hillman (1981; 1984) and Glynis Jones (1984; 1987). More recently still, some archaeo- and ethno-botanists have sought insights into the domestication process from the varied ethnographically recorded roles of plants in human social relations, as gifts, festal foods, territorial markers, symbols of group identity, and in other ways (see, e.g., Dietler & Hayden 2001; Hastorf 1998; 2004; Ingold 1996; Rival 1998). This emphasis on the multifaceted social functions of plants in human affairs increases our awareness of the complexities of early plant use, but it also makes it even more difficult to link the usually exiguous archaeobotanical evidence directly to the ethnographic record.

This task is even more challenging in archaeological studies of the role of plants in the lives of former foragers than in those of past agriculturalists, because most archaeological investigations of hunter-fisher-gatherers are concerned with more distant (Palaeolithic and Mesolithic) time periods than studies of (Neolithic and later) farmers. The examples presented here that combine ethnohistorical, ethnobotanical, and archaeological study of recent foragers in the Queensland rain forest and Torres Strait Islands demonstrate how much more feasible and productive it is to investigate past forager subsistence when the time gap between ethnographically observed and archaeologically inferred subsistence practices is measured in centuries rather than millennia. But, despite the greater difficulty inherent in investigating plant use by people who lived many millennia ago, by combining the use of new analytical techniques with more refined models of plant-people interaction based on ethnographic and ecological data, we can expect to achieve great gains in our understanding of the myriad ways in which humans have procured and produced food from plants, in the tropics and elsewhere.

REFERENCES

ALLEN, H. 1974. The Bagundji of the Darling Basin: cereal gatherers in an uncertain environment. *World Archaeology* 5, 309-22.

BAILEY, F.M. 1899-1905. *The Queensland flora*. 7 vols. Brisbane: Queensland Government.

―――― 1909. *Comprehensive catalogue of Queensland plants both indigenous and naturalized*. (Second edition). Brisbane: Queensland Government.

BARHAM, A.J. & D.R. HARRIS 1985. Relict field systems in the Torres Strait region. In *Prehistoric intensive agriculture in the tropics*, part 1 (ed.) I.S. Farrington, 247-283. Oxford: British Archaeological Reports International Series 232.

BARKER, G. 2002. Prehistoric foragers and farmers in South-East Asia: renewed investigations at Niah Cave, Sarawak. *Proceedings of the Prehistoric Society* 68, 147-64.

BARTON, H. 2005. The case for rain forest foragers: the starch record at Niah cave, Sarawak. In *The human use of caves in peninsular and island southeast Asia* (eds) G. Barker & D. Gilbertson. (Special Issue). *Asian Perspectives* 44, 56-72.

―――― & V. PAZ in press. Subterranean diets in the tropical rain forests of Sarawak, Malaysis. In *Rethinking agriculture: archaeological and ethnoarchaeological perspectives* (eds) T.P. Denham, J. Iriarte & L. Vrydaghs. London: UCL Press.

BOWDERY, D. 1999. Phytoliths from tropical sediments: reports from Southeast Asia and Papua New Guinea. *Bulletin of the Indo-Pacific Prehistory Association* 18, 159-68.

BURKILL, I.H. 1966. *A dictionary of the economic products of the Malay Peninsula*. 2 vols. Kuala Lumpur: Crown Agents for the Colonies.

CANE, S. 1989. Australian Aboriginal seed grinding and its archaeological record: a case study from the Western Desert. In *Foraging and farming: the evolution of plant exploitation* (eds) D.R. Harris & G.C. Hillman, 99-119. London: Unwin Hyman.

DENHAM, T., S. HABERLE & C. LENTFER 2004. New evidence and revised interpretations of early agriculture in Highland New Guinea. *Antiquity* 78, 839-57.

――――, ――――, ――――, R. FULLAGAR, J. FIELD, M. THERIN, N. PORCH & B. WINSBOROUGH 2003. Origins of agriculture at Kuk Swamp in the highlands of New Guinea. *Science* 301, 189-93.

DIETLER, M. & B. HAYDEN (eds) 2001. *Feasts: archaeological and ethnographic perspectives on food, politics and power*. Washington, D.C.: Smithsonian Institution Press.

FLECKER, H.G.B., S. STEPHENS & S.E. STEPHENS 1948. *Edible plants in North Queensland*. Cairns: North Queensland Naturalists Club.

FULLER, D.Q., R. KORISETTAR, P.C. VENKATASUBBAIAH & M.K. JONES 2004. Early plant domestications in southern India: some preliminary archaeobotanical results. *Vegetation History and Archaeobotany* 13, 115-29.

GOLSON, J. 1997. From horticulture to agriculture in the New Guinea highlands. In *Historical ecology in the Pacific islands: prehistoric environmental and landscape change* (eds) P.V. Kirch & T.L. Hunt, 39-50. New Haven: Yale University Press.

———, R.J. LAMPERT, J.M. WHEELER & W.R. AMBROSE 1967. A note on carbon dates for horticulture in the New Guinea Highlands. *Journal of the Polynesian Society* 76, 369-71.

GOWLETT, J.A.J. & R.E.M. HEDGES (eds) 1986. *Archaeological results from accelerator dating*. Oxford: Oxford University Committee for Archaeology, Monograph 11.

HALE, H.M. & N.B. TINDALE 1933. Aborigines of Princess Charlotte Bay, north Queensland. *Records of the South Australia Museum* 5, 63-116.

HALLAM, S. 1989. Plant usage and management in southwest Australian Aboriginal societies. In *Foraging and farming: the evolution of plant exploitation* (eds) D.R. Harris & G.C. Hillman, 136-51. London: Unwin Hyman.

HARRIS, D.R. 1971. The ecology of swidden cultivation in the upper Orinoco rain forest, Venezuela. *Geographical Review* 61, 475-95.

——— 1972. The origins of agriculture in the tropics. *American Scientist* 60, 180-93.

——— 1973. The prehistory of tropical agriculture: an ethnoecological model. In *The explanation of culture change: models on prehistory* (ed) C. Renfrew, 391-417. London: Duckworth.

——— 1977. Subsistence strategies across Torres Strait. In *Sunda and Sahul: prehistoric studies in South-east Asia, Melanesia and Australia* (eds) J. Allen, J. Golson & R. Jones, 421-63. London: Academic Press.

——— 1978. Adaptation to a tropical rain-forest environment: Aboriginal subsistence in northeastern Queensland. In *Human behaviour and adaptation* (eds) N. Blurton-Jones & V. Reynolds, 113-34. London: Taylor & Francis.

——— 1979. Foragers and farmers in the western Torres Strait Islands: an historical analysis of economic, demographic, and spatial differentiation. In *Social and ecological systems* (eds) P.C. Burnham & R.F. Ellen, 75-109. London: Academic Press.

——— 1984. Ethnohistorical evidence for the exploitation of wild grasses and forbs: its scope and archaeological implications. In *Plants and ancient man: studies in palaeoethnobotany* (eds) W. van Zeist & W.A. Casparie, 63-9. Rotterdam: Balkema.

——— 1987a. The impact on archaeology of radiocarbon dating by accelerator mass spectrometry. *Philosophical Transactions of the Royal Society London A* 323, 23-43.

——— 1987b. Aboriginal subsistence in a tropical rain forest environment: food procurement, cannibalism and population regulation in northeastern Australia. In *Food and evolution: toward a theory of human food habits* (eds) M. Harris & E.B. Ross, 357-85. Philadelphia: Temple University Press.

——— 1996. Domesticatory relationships of people, plants and animals. In *Redefining nature: ecology, culture and domestication* (eds) R. Ellen & K. Fukui, 437-63. Oxford: Berg.

HASTORF, C.A. 1998. The cultural life of early domestic plant use. *Antiquity* 72, 773-82.

——— 2004. Plant domestication: making society and marking territory with food. In *Cultural diversity and the archaeology of the 21st century*, 24-31. Okayama: Society of Archaeological Studies, 50th Anniversary Symposium.

HATHER, J.G. 1988. The anatomical and morphological interpretation and identification of charred parenchymatous plant tissues. Unpublished Ph.D. thesis, University of London.

―――― 1991. The identification of charred archaeological remains of vegetative parenchymatous tissues. *Journal of Archaeological Science* **18**, 661-75.

―――― 1994. The identification of charred root and tuber crops from archaeological sites in the Pacific. In *Tropical archaeobotany: applications and new developments* (ed.) J.G. Hather, 51-64. London: Routledge.

―――― 2000. *Archaeological parenchyma*. London: Archetype.

―――― & N. HAMMOND 1994. Ancient Maya subsistence diversity: root and tuber remains from Cuello, Belize. *Antiquity* **68**, 330-5.

―――― & P.V. KIRCH 1991. Prehistoric sweet potato (*Ipomoea batatas*) from Mangaia Island, central Polynesia. *Antiquity* **65**, 887-93.

HILLMAN, G.C. 1981. Reconstructing crop husbandry practices from charred remains of crops. In *Farming practice in British prehistory* (ed.) R. Mercer, 123-62. Edinburgh: University Press.

―――― 1984. Interpretation of archaeological plant remains: the application of ethnographic models from Turkey. In *Plants and ancient man: studies in palaeoethnobotany* (eds) W. van Zeist & W.A. Casparie, 1-42. Rotterdam: Balkema.

INGOLD, T. 1996. Growing plants and raising animals: an anthropological perspective on domestication. In *The origins and spread of agriculture and pastoralism in Eurasia* (ed.) D.R. Harris, 12-24. London: UCL Press; Washington, D.C.: Smithsonian Institution Press.

JONES, G.E.M. 1984. Interpretation of archaeological plant remains: ethnographic models from Greece. In *Plants and ancient man: studies in palaeoethnobotany* (eds) W. van Zeist & W.A. Casparie, 43-61. Rotterdam: Balkema.

―――― 1987. A statistical approach to the archaeological identification of crop processing. *Journal of Archaeological Science* **14**, 311-23.

JONES, R. & B. MEEHAN 1989. Plant foods of the Gidjingali: ethnographic and archaeological perspectives from northern Australia on tuber and seed exploitation. In *Foraging and farming: the evolution of plant exploitation* (eds) D.R. Harris & G.C. Hillman, 120-35. London: Unwin Hyman.

LAMPERT, R.J. 1967. Horticulture in the New Guinea Highlands – C14 dating. *Antiquity* **41**, 307-9.

LATHRAP, D.W. 1977. Our father the cayman, our mother the gourd: Spinden revisited, or a unitary model for the emergence of agriculture in the New World. In *Origins of agriculture* (ed.) C.A. Reed, 713-51. The Hague: Mouton.

LAWTON, H.W., P.J. WILKE, M. DEDECKER & W.M. MASON 1976. Agriculture among the Paiute of Owens Valley. *Journal of California Anthropology* **3**, 13-50.

LEBOT, V. 1999. Biomolecular evidence for plant domestication in Sahul. *Genetic Resources and Crop Evolution* **46**, 619-28.

LOY, T.H. 1994. Methods in the analysis of starch residues on prehistoric stone tools. In *Tropical archaeobotany: applications and new developments* (eds) J.G. Hather, 86-114. London: Routledge.

MACGILLIVRAY, J. 1852. *Narrative of the voyage of HMS 'Rattlesnake' ... during the years 1846-1850 ...* , vol. 2. London: Boone.

MANNINO, M. & K. THOMAS 2003/4. A site for all seasons? Prehistoric coastal subsistence in northwest Sicily. *Archaeology International*, 31-4.

MBIDA, C.M., H. DOUTRELEPONT, L. VRYDAGHS, R.L. SWENNAN, R.J. SWENNAN, H. BEECKMAN, E. DE LANGHE & P. DE MARET 2001. First archaeological evidence of banana cultivation in central Africa during the third millennium before present. *Vegetation History and Archaeobotany* **10**, 1-6.

――――, ――――, ――――, ――――, ―――――――― & ―――― 2004. Yes there were bananas in Cameroon more than 2000 years ago. *InfoMusa* **13**, 40-2.

———, W. VAN NEER & L. VRYDAGHS 2000. Evidence for banana cultivation and animal husbandry during the first millennium BC in the forest of southern Cameroon. *Journal of Archaeological Science* 27, 151-62.

MESTON, A. 1904. *Report on the expedition to the Bellenden-Ker Range.* Brisbane: Queensland Government.

MOORE, D.R. 1978. *Islanders and Aborigines at Cape York*, Canberra: Australian Institute of Aboriginal Studies.

O'CONNELL, J.F., P.K. LATZ & P. BARNETT 1983. Traditional and modern plant use among the Alyawara of central Australia. *Economic Botany* 37, 80-109.

PARFITT, S.A. & M.B. ROBERTS 1999. Human modification of faunal remains. In *Boxgrove: a middle Pleistocene hominid site at Eartham Quarry, Boxgrove, West Sussex* (eds) M.B. Roberts & S.A. Parfitt, 395-419. London: English Heritage.

PEARSALL, D.M. 1994. Phytolith analysis. In *Regional archaeology in northern Manabi, Ecuador*, vol. 1: *Environment, cultural chronology, and prehistoric subsistence in the Jama River Valley* (eds) J.A. Ziedler & D.M. Pearsall, 161-74. Pittsburgh: Department of Anthropology, University of Pittsburgh.

PERRY, L. 2002. Starch granule size and the domestication of manioc (*Manihot esculenta*) and sweet potato (*Ipomoea batatas*). *Economic Botany* 56, 335-49.

——— 2004. Starch analyses reveal the relationship between tool type and function: an example from the Orinoco valley of Venezuela. *Journal of Archaeological Science* 31, 1069-81.

PIPERNO, D.R. 1988. *Phytolith analysis: an archaeological and geological perspective.* San Diego: Academic Press.

——— 1989. Non-affluent foragers: resource availability, seasonal shortages, and the emergence of agriculture in Panamanian tropical forests. In *Foraging and farming: the evolution of plant exploitation* (eds) D.R. Harris & G.C. Hillman, 538-54. London: Unwin Hyman.

——— & I. HOLST 1998. The presence of starch grains on prehistoric tools from the humid neotropics: indications of early tuber use and agriculture in Panama. *Journal of Archaeological Science* 25, 765-76.

———, K. HUSUM CLARY, R.G. COOKE, A.J. RANERE & D.W. WEILAND 1985. Preceramic maize from central Panama: evidence from phytoliths and pollen. *American Anthropologist* 87, 871-8.

——— & D.M. PEARSALL 1998. *The origins of agriculture in the lowland neotropics.* San Diego: Academic Press.

———, A.J. RANERE, I. HOLST & P. HANSELL 2000. Starch grains reveal early root crop horticulture in the Panamanian tropical forest. *Nature* 407, 894-97.

RIVAL, L. 1998. Domestication as a historical and symbolic process: wild gardens and cultivated forests in the Ecuadorian Amazon. In *Advances in historical ecology* (ed.) W. Balée, 232-50. New York: University of Columbia Press.

ROSENDAHL, P. & D.E. YEN 1971. Fossil sweet potato remains from Hawai'i. *Journal of the Polynesian Society* 80, 379-85.

SAUER, C.O. 1936. American agricultural origins: a consideration of nature and culture. In *Essays in anthropology presented to A.L. Kroeber in celebration of his sixtieth birthday, June 11, 1936* (ed.) R.H. Lowie, 278-97. Berkeley: University of California Press.

——— 1952. *Agricultural origins and dispersals.* New York: American Geographical Society.

SIMMONDS, N.W. 1976. Bananas. In *Evolution of crop plants* (ed.) N.W. Simmonds, 211-15. London: Longman.

SMITH, B.D. 2001. Low-level food production. *Journal of Archaeological Research* 9, 1-43.

STEWARD, J.H. 1930. Irrigation without agriculture. *Papers of the Michigan Academy of Sciences, Arts and Letters* 12, 149-56.

THOMAS, K.D. & M.A. MANNINO 2003. Archeomalacologia della Sicilia nord-occidentale: un programma di ricerca per lo studio dell'ecologia e della sussistenza umana nella preistoria tramite l'analisi dei gusci di molluschi marini da siti archeologici. *Quaderni del Museo Archeologico Regionale 'Antonio Salinas'* **7**, 45-58.

UGENT, D., S. POZORSKI & T. POZORSKI 1982. Archaeological potato tuber remains from the Casma valley of Peru. *Economic Botany* **36**, 182-92.

———, ——— & ——— 1986. Archaeological manioc (*Manihot*) from coastal Peru. *Economic Botany* **40**, 78-102.

VANSINA, J. 2004. Banana in Cameroon *c*. 500 BCE? Not proven. *Azania* **38**, 174-6.

WILSON, S.M. 1985. Phytolith evidence from Kuk, an early agricultural site in New Guinea. *Archaeology in Oceania* **20**, 90-7.

5
Amazonian historical ecologies

LAURA RIVAL

My purpose in this chapter is to illustrate the theoretical power of historical ecology as a holistic approach to human ecology which, particularly in the Amazonian context, has opened new paths for the integration of the cognitive, historical, and political dimensions of human nature. The greatest strength of historical ecology, in my view, is that it allows us to formulate non-reductionist analyses of the environment/society interface. I start with a brief exposition of Julian Steward's ideas to show that if they have – unfortunately – given rise to a reductionist model of functional adaptation according to cost/benefit criteria, they can also be interpreted in a way that makes him the unmistakable and direct precursor of the historical ecology approach. I then discuss Bill Balée's model of agricultural regression to show why and how botanical and ethnobotanical data have come to play a key role in Amazonian anthropological research since the early 1990s. Given the extraordinary demographic changes that have occurred in Amazonian societies at various points in time, but especially in the aftermath of the Spanish and Portuguese Conquest, it would have been impossible to re-assess the historical trajectory of these societies without botanical and ethnobotanical studies. In the last part of the paper, I discuss critically the thesis of post-contact cultural regression. I examine the ways in which historical ecology has been used to research nomadic bands subsisting with few or no domesticates in lowland South America. I argue that trekking, far from representing a necessary intermediary stage in the regression from horticulture to foraging, constitutes, in some cases, a *sui generis* solution to deep contradictory forces of a political, religious, and social character. Such internal processes may have long predated the Conquest and the disruptions it caused.

Historical ecology: origins and principles

Bill Balée and Caroline Crumley, chief advocates of the historical ecology perspective, have shown that it is theoretically continuous with Steward's cultural ecology. They define historical ecology as the undertaking of a diachronic analysis of living ecological systems, with the view to account fully for their structural and functional properties. In their various publications, they have explained how nature is always in the making, and why human-environment interaction should be analysed from a historical perspective. Crumley, who traces the origin of the term 'historical ecology' to the archaeological palynologist Edward Deevey's research programme at the University of Florida in the early 1970s, stresses the fact that the physical world and human societies are always and everywhere inextricably linked (Crumley 1998: vi). Everywhere, human action shapes, and is shaped by, the physical environment, without being necessarily destructive. Balée argues that historical ecology is 'even more dialectical than historical materialism, for it begins with the dialectic of an inalienable link between nature and culture' (1998: 4); and more holistic, for it attempts to understand the mutual influence of people and non-human nature, a process which is at once biological and cultural. Historical ecology, more an approach or a research strategy than a paradigm, addresses a central question: 'How does environmental change relate to the historical development of human societies?' An integral part of the new ecological anthropology, historical ecology seeks to reconceptualize the complexity of the biological world, particularly the problematic distinction between the wild and the domesticated, which has inspired natural science research on the diversity of biological life. The way in which society views nature is in part a function of how society has affected nature. Nature and the cultural conceptions of nature develop together; they co-evolve.

Wrongly characterized by some as single-minded environmental determinism, Julian Steward's position was premised on the fact that the superorganic factor of culture affects, and is affected by, the total web of life. Steward, who did not consider culture to form a seamless whole, tried to introduce a priority order among the various cultural components. He saw the way people work as the determining factor in cultural adaptation. Taking ecology in its broadest meaning, he focused his analytical effort on subsistence activities and their influence upon a local society's material expressions and values. Although he especially emphasized the relationship between the tools that humans use in pursuit of subsistence and certain environmental features,

Steward did not ignore worldviews, but examined them in the light of historical contingencies.

Steward was primarily concerned with social change. His framework was materialist, historical, and evolutionary, not only because he understood cultural-ecological adaptation with reference to both culture and the environment, but also because he looked at the effect of the environment upon culture in terms of ecological *and* historical processes. His intent to reconcile diachronic and synchronic explanations of cultural facts in his search for laws of social transformation was truly original at the time, and remains an inspiration today. One can best appreciate this by comparing his theory with the equally innovative dynamic theory of institutional change proposed by Edmund Leach in his celebrated *Political systems of highland Burma* (1954). Whereas both Leach and Steward asserted that relations between people do not change durably before their relationship to the natural environment is undermined and requires new forms of adaptation, only Steward dedicated his research to the precise environmental and historical conditions underlying cultural adaptation. He would not have mistaken the Kachin for traditional hill farmers, and he would not, as Leach did, have ignored the fact that the Kachin were intensively involved in poppy production, tax levy, and long-distance trade.

When linking subsistence and social structure, or arguing that subsistence plays a role in determining social and political boundaries, Steward always emphasized local processes of adaptation. One can argue that, for him, habitat becomes the driving force for social and cultural change only for those who possess accurate and encyclopaedic knowledge of local environmental conditions. The application of ecological functionalism to Amazonia produced poor analytical results because it was narrowly focused on adaptation to scarce environmental resources. Our current knowledge of the Amazon biome is far more sophisticated than it was in the late 1940s and early 1950s, when Steward edited the *Handbook of South American Indians* (1946). This new knowledge allows us to rethink human occupation and adaptation to the Amazon, to redefine the forces that shape the material dimensions of social life, and to recognize that Amazonian hunter-gatherers have played an active part in the making of the environment that they have occupied for millennia (Rival 1999).

Anthropogenic forest formations in Amazonia[1]

Since the pioneering work summarized in Steward (1946), many cultural ecology theories have been proposed to explain prehistorical subsistence patterns and historical agricultural techniques in Amazonia. Betty Meggers's (1996 [1971]) environmental determinist thesis was followed by works bent on proving that contemporary native Amazonia is the result of severe environmental limitations. Low population density, incipient warfare, transient slash-and-burn horticulture, and food taboos have, for instance, been analysed as manifestations of human adaptation to natural resource scarcity (Rival 1999).

It is through careful examination of botanical and ethnobotanical data that authors such as Posey and Balée (Balée 1993; Posey 1984; Posey & Balée 1989) have marshalled new empirical evidence to challenge the earlier claims that Amazonia lacked in resource potential, or that human activity in this part of the world was no more than mere disturbance of natural processes. Posey (1984) argued that the Kayapó Indians are not simple cultivators who periodically cut the forest down in order to open new gardens. Rather, they practise a form of 'agroforestry', by which he meant that the limited, shifting, and periodic removal of the forest cover to cultivate foodcrops is only one phase of a long cycle of integrated forest management. Balée (1989; 1992; 1993) has similarly argued that the distribution of tree species in the Amazon rain forest is deeply influenced by human intentional and non-intentional management. Local concentrations of specific plant species, which are good indicators of past human activity, show as empirically as any other reliable historiographic record would that environment and society interacted historically in Amazonia. Balée's study of the indigenous agroforestry complex in Maranhão, Brazil, for example, explicitly defends the idea that forests of biocultural origin can be treated as objective records of past human interactions with plants, even if the local population does not have any social memory of such history and cannot differentiate old fallows from patches of undisturbed forest (Balée 1993). The presence of surface pottery and charcoal in the soil, the distribution of species, the size of trunks, oral history, and native classifications of forest and swidden types can all be used to differentiate old fallows from high forests. This leads Balée to argue that nomadic bands do not wander at random, but move their camps between palm forests, bamboo forests, or Brazil nut forests, which are all 'cultural forests', that is, ancient dwelling sites. Amazonian foraging bands such as the Guajá, the Kaingang, or the Sirionó are able to subsist in the rain forest without cultivated crops, on the basis of a few essential 'wild' resources (palms, fruit trees, or bamboo), which are, in fact, the products of the activities of ancient populations. According to Balée, these groups regressed in the aftermath of the Conquest from village life and cultivation to trekking in old fallows and foraging in managed forests created by pre-Columbian societies. They lost most of their domesticated crops in the process. Evidence from observation of contemporary gardening activities; the wide occurrence of charcoal and numerous potsherds in the forest soil; the greater concentration of palms, lianas, fruit trees, and other heavily used forest resources on archaeological sites, as well as induction about the long-lasting effects of past human interference, have all led Balée to argue that, far from having been limited by scarce resources, the indigenous people of Amazonia have created biotic niches or 'anthropogenic forests' since prehistoric times; these have been exploited continuously to this day. Like many contemporary analysts, and most especially the archaeologist Anna Roosevelt (1991; 1993; 1994; 1998), the geographer William Denevan (1996; 2001), and the botanist Charles Clement (1999), Balée stresses the significance of post-Columbian demographic collapse. Before this historical watershed, Amazonia was peopled with complex societies made up of large semi-sedentary villages subsisting on a wide range of intensively cultivated domesticated and semi-domesticated crops. These settlements were networked across impressive distances through elaborate exchange and trade routes. Like these authors, Balée (1999: 26) stresses the dramatic loss of social, political, and techno-economic complexity in the majority of native societies in modern times.

Balée and Posey's work has been extremely influential in Amazonian anthropology, and beyond. Their research has shown that adaptation is a dynamic process in which human agency plays a central role. The environment does not limit cultural development as extensively as was previously believed; and what looks like a pristine environment may in fact turn out to be an ancient agricultural site. Balée and Posey show that native populations modified their surroundings through a variety of activities that ultimately transformed the Amazonian landscape in distinctly human ways. They were also the first authors to emphasize that human activity, in addition to enriching the vegetation cover and diversifying plant species, also improved soils (see Lehman, Kern, Glaser & Woods 2003). Three lessons can thus be learned from their work. On the one hand, human populations must be treated as full components of the ecosystems in which they live; on the other hand, equal importance must be granted to the history of ecological systems and to that of human societies. Finally, Balée and Posey demonstrate the relevance of botanical and ethnoecological studies for establishing the link between the physical and social past in Amazonia.

Building on the research of Balée, Posey, and others, Clement (1999) offers the most comprehensive synthesis of Amazonian crop genetic biogeography to date. In so doing, he goes a step further in arguing for the full integration of environmental and historical explanations. The goal this time is not so much to estimate what has been lost since 1492, but to hypothesize how much agrobiodiversity was created by humans in the Amazon region between the late Pleistocene and the fifteenth century. Clement applies the new evolutionary theory developed by David Harris (e.g. Harris 1996; Harris & Hillman 1989) and David Rindos (1984) during the 1980s to evaluate the rich and varied pre-Columbian crop genetic heritage. At least 257 plant species were cultivated in the Americas in the fifteenth century (Clement 1999: 188), of which 79 probably originated in the Amazon basin, and another 37 in neighbouring regions (Clement 1999: 203). Some 138 crops belonging to 44 botanical families were either cultivated or more or less directly affected by human management in Amazonia at the time of contact (Clement 1999: 192). The distinction between species and landscape domestication, as well as the integration of both processes, is central to Clement's reconstruction of Amazonian horticulture. Using Harris's definition of domestication as a co-evolutionary process based on a continuum of human investment in selection and environmental manipulation (Clement 1999: 189), the author proposes a classification of Amazonian cultivated plants according to their degree of domestication (full domesticates, semi-domesticates, and incipient domesticates) and their particular life history (annuals, semi-annuals, and perennials). Six categories of plants are distinguished on the basis of these variables, which allow Clement to reconstruct the inter-related historical ecology of anthropogenic forest formations and of crop genetic resources. He concludes that if a high percentage (68%) of Amazonian domesticates, semi-domesticates, and incipient domesticates are trees and woody perennials, this fact is to be attributed primarily not to the nature of the forest ecosystem but, rather, to the high dependence of domesticated annuals on human management. Crop genetic erosion paralleled the physical disappearing of human populations. This explains why diversity, especially infraspecific diversity of cultivars, was reduced shortly after large indigenous Amazonian societies succumbed to world diseases and depopulation.

The (agri)cultural regression model

In the previous section, I have presented recent studies of tree species distribution and intra-species diversity which show conclusively that indigenous peoples did not passively adapt to a forest environment poor in vital resources. On the contrary, Amazonian peoples actively manipulated the forest ecosystem, enriched the soils, managed and diversified a wide range of plant species, and generally created the material and physical conditions to maintain surprisingly high levels of population density, at least in some areas. In this section, I examine the use of botanical and ethnobotanical data in works that elaborate on the losses incurred upon contact.

As mentioned earlier, most current authors explain the presence of foraging and trekking populations in Amazonia as a post-Conquest phenomenon. Balée's work on Brazilian marginal groups has led him to conclude that Europe's invasion of lowland South America caused not only the demographic collapse of native populations, but also their massive cultural devolution, with surviving societies adapting to biocultural forests resulting from the dynamic interaction between history and ecology (Balée 1992). It is only because they have secured access to a few essential non-domesticates such as palms and other fruit trees (which, in fact, result from the horticultural activities of ancient populations) that Amazonian foragers have been able to survive without cultivated crops. The process of agricultural regression, parallel to that of regression from village life to full nomadism, is progressive. At each stage, a cultigen is lost, and dependence on uncultivated plants increases. The originality of Balée's argument is to show that loss of cultigens and increased reliance on uncultivated plants amount not to a return to pristine nature, but, instead, to a process of adaptation to 'vegetational artefacts of another society' (Balée 1988: 48). Moreover, in his view, the botanical knowledge of deculturated foragers, that is, their knowledge of both garden crops and wild plants, is poorer than that of gardeners. Present-day foragers who do not remember that their forebears cultivated have forgotten the technical *savoir-faire* of their predecessors, but their languages still possess cognates for cultigens. Said differently, the only cultural transmission that has successfully operated through time is unconscious linguistic knowledge. However, even linguistic knowledge may be erased over time. Balée (1992) mentions, for instance, that the Guajá, who still have a cognate for maize, have lost the term for bitter manioc.

Balée's twin use of linguistic and botanical data to infer past agricultural patterns and historical change recalls Sapir's (1912) study of diachronic change in folk-botanical terminology. Sapir thought that human languages encode memories of human inventiveness, adaptation, and survival skills, and co-evolve with environmental knowledge. The linguist Louisa Maffi has also used Sapir's theory to show that 'the vocabulary of a language ... inventories the characteristics of the local flora and fauna and thus bears the stamp of the physical environment in which the speakers are placed' (2001: 25). This is why Maffi correlates acculturation (erosion of both native language and ecological knowledge) with losses in biological diversity. She therefore proposes a model of regression akin to that proposed by Clement, who correlates demographic collapse with genetic erosion, or by Balée, who correlates colonial expansion with agricultural regression. In all these scenarios, historical events disrupt cultural developments, and changes in socioeconomic systems have profound ecological as well as cultural consequences.[2]

I agree with these authors that human/environment interactions must be analysed holistically, that is, by taking into consideration both the physical and the mental world – in this case botanical knowledge and the vocabulary used to encode it. However, the inextricable link between the physical world and human societies need not be understood according to evolutionary/devolutionary processes, which ultimately imply that agricultural intensification is inherently progressive.[3] Moreover, a society's mental world encompasses far more than botanical terms. Let us examine the agricultural regression thesis critically, starting with Balée's arguments.

For Balée (1995), history is far more relevant than evolution to understand the changes that occurred in the relationship between human societies and their environments, but what he really means by history is indigenous adaptation to the Spanish Conquest and to post-Conquest biological and political dynamics. History in his model is never envisaged as resulting from pre-Conquest social contradictions or political conflicts; it is always a reaction to external events which invariably affect developmental and evolutionary trends 'backwards': that is, by reversing the pace of development. The primary historical constraint faced by native Amazonians after the Conquest was the severe demographic collapse that they experienced. All other regressive changes are seen as consequential: the loss of agricultural knowledge due to defective cultural transmission, and the regression from agriculture to trekking to pure foraging, because foraging is the only survival option available to small bands.[4] Balée therefore correlates lack of agriculture and mobility. People who farm intensively are not highly mobile; conversely, people who hunt and gather are highly mobile. While I agree with Balée's insistence that mobility be considered as an adaptive strategy to historical, rather than to environmental, conditions, I lament the fact that he overlooks pre-Conquest historical dynamics triggered by conflicts between highly mobile and less mobile native populations.

Another critical aspect of Balée's model is that it is still cast in the mould of optimal foraging. What Balée really objects to in the work of optimal foraging theorists is not that maximization of benefits or minimization of costs is at the root of Amazonian subsistence strategies, but, rather, that optimal foraging theory fails to recognize that trekking and foraging may be adaptations to cultural rather than to natural environments. What differentiates the adaptation of Amazonian trekkers and foragers is that, although they depend on domesticated and cultivated crops, they need neither to cultivate garden products, nor to exchange game or collected forest products for garden products. The resources they gather in the wild exist as the result of the activities of past agriculturalists, who have modified the forest environment to the point at which it can be described no longer as a natural habitat, but, rather, as a cultural landscape.[5] In this sense, Balée's view of human adaptation is not substantially different from that of Kim Hill (K. Hill & Hurtado 1996), for instance, who analyses the mobility of Aché foragers as a function of their adaptation to a particular type of biocultural landscape. Trekking and foraging as envisaged by Balée can still be analysed as a form of economic adaptation to a particular environment; in this case, former agriculturalists such as the Aché have maximized their adaptation by adapting the environment to their needs, rather than adapting their needs to the environment. In other words, there is no room in Balée's model for understanding the subsistence activities of trekkers and foragers in cultural terms, that is to say, for including in the analysis their own conceptualization of gathering and hunting in cultural landscapes, or their own discourse about their subsistence practices. We need to explore whether groups who are known to have been intensive horticulturalists in the past and who now mainly forage, hunt, and gather do so similarly to those who are still intensive horticulturalists, and whether their symbolic representations of hunting, gardening, and foraging are identical to those heralded by intensive horticulturalists. In short, does hunting and gathering in old fallows, that is, in environments modified by previous human intervention and management, make a difference, practically and symbolically?

Balée's main concern is to understand the shift from agriculture to foraging in Amazonian societies that are found today on archaeological sites which were once inhabited by ancient chiefdoms for which agriculture must have played a central role. He shares this concern with a number of authors, in particular Roosevelt (1993), Lathrap (1970; 1973), and, to some extent, Lévi-Strauss (1948; 1955; 1963*a*; 1963*b*). Like these authors, because he sees the progressive abandonment of village life and horticulture by a number of Tupi-Guarani, the intensive horticultural systems of whom we know from ethnohistorical sources were destroyed by warfare and epidemics, he assumes that Amazonian societies are fundamentally of Denevan's 'intensive horticulture' type. However, Denevan (1996; 2001) stresses that the hinterlands were *simultaneously* used by indigenous populations living in sedentary, densely populated village settlements, and by small, mobile groups dispersed throughout the forest. Denevan's thesis may be used to support a view of the dynamic history of plant/human interaction according to which foraging with incipient horticulture is as much a cultural choice as intensive, sedentary agriculture.[6] Denevan's work thus offers an important corrective to the thesis proposed by Clement, Balée, Roosevelt, and other authors who implicitly conclude that, if it had not been for the Conquest, Amazonia's native populations would have continued to develop intensive agriculture and would have increasingly complexified.

To summarize, the agricultural regression thesis defended by Balée is unsatisfactory on a number of counts. The first problem with this thesis is that it only considers post-Conquest historical causes, and never pre-Conquest ones. In other words, devolution to a simpler, less hierarchical social form and to a more nomadic, less agriculture-dependent way of life is envisaged only as the outcome of European expansion. The fact that there might have been processes of regression from sedentism to nomadism and loss of cultigens *before* the Iberian Conquest is simply not envisaged. Consequently, intensive, sedentary agriculture is premised as being logically and historically anterior to the forms of trekking and foraging observed today. And whereas the sequence foraging → sedentary horticulture is naturalized as evolutionary, thus de-historicized, the sequence sedentary horticulture → foraging is interpreted as having been induced by the historical encounter between Europe and pre-historical Amazonia. Like much post-colonial historico-anthropological writing which assumes that history started with the Iberian arrival (J. Hill & Santos Granero 2002), this position impoverishes the actual diversity of historical trajectories in the region. It has been my contention that nomadism and a hunting-gathering way of life need not be post-colonial, for they may represent cultural and political choices already present in pre-Conquest values and social forms.

Furthermore, the devolution thesis fails to explain why *both* sedentarized horticulture (as practised, for example, by Quichua communities) and trekking (as practised, for instance, by Huaorani bands) represent successful adaptations to the Upper Napo ecology. The extraction of wild resources through hunting and gathering, although qualitatively different from the use of domesticates through cultivation, nevertheless affects species distribution, and, consequently, systematically modifies the environment (Yen 1989). Finally, trekkers who continue to rely on non-cultivated food develop forest management practices which lead to greater concentrations of favoured resources within specific areas. In so doing, they do transform nature, albeit in a distinctive way, for their techniques are not geared to intensify production outputs.

Another and related problem with the devolution thesis is that it overemphasizes the evolutionary significance of domestication, and treats swidden horticulture as a homogeneous and empirical category, ignoring the importance of subsistence modes in defining group identity. Foragers and sedentarized cultivators have developed radically different means of associating with plants and alternative ways of being in the world and of knowing it. Rather than a continuum of intensity of resource exploitation, foraging and cultivation constitute alternative strategies of resource procurement and modes of practical and intentional engagement with their environment. This difference is significant, particularly in terms of identity formation and inter-ethnic relations, given that, on the whole, gardeners feel morally superior to foragers and trekkers (Rival 1999: 82). Last but not least, the devolution thesis entirely ignores the fact that trekkers and foragers do not represent themselves as less developed than cultivators. Symbolic constructions of ecologies of difference must therefore be taken into consideration; they form an essential part of the analysis. Culture is a way of life, and identity as forager or as cultivator an important cultural marker. The fact that the Guajá, for instance, are fond of saying that they do not cultivate, but raid instead, illustrates the fact that to hunt and to gather represents far more than economic action. What is being expressed here are cultural and political values that need to be understood as such.

In his paper in this volume, David Harris comments on the difficulties of assessing the archaeological data from the Amazon basin and other tropical sites, and on the disparity between the picture for early plant domestication emerging from ethnography and that reconstituted by archaeology. In addition, we might mention the disagreements existing between archaeologists on the existence and the form of pre-Columbian chiefdoms. Archaeologists and anthropologists working in the cultural ecology tradition stress the social and cultural discontinuity between pre-Columbian and contemporary Amazonian societies, with their basic social organization of small, politically independent and egalitarian local groups formed through cognate ties. They treat high mobility and foraging as indicators of historical change. The nomadic, foraging way of life of interfluvial groups does not, they argue, reflect the pattern that predominated in pre-Columbian Amazonia, where elaborate autochthonous chiefdoms developed and flourished. What is at stake in this debate is the nature of the changes that occurred before, during, and after the Conquest. How are we best to understand the interactions between the natural history of the forest and the eventful, uneven, and violent history of human societies in this part of the world?

To begin to offer plausible answers to this question, I wish to argue, we need to consider biocultural data as more than a repository of historical evidence. We need to describe ways of life ethnographically, and this necessarily implies analysing Amazonian hunter-gatherers' own social and symbolic structuring of the world. We need to establish whether a particular socio-cultural fact is continuous or discontinuous with the past as it is both historiographically recorded and culturally encoded. Balée's contribution has been to highlight that higher mobility in Amazonia is linked to the choice of using resources that are not 'wild' but 'biocultural', as well as to the choice of replacing cultivation with gathering. We now need to determine how such shifts in social and economic practices are reflected at the level of collective representations and in the formation of distinctive identities.

The contrastive historical ecologies of Amazonian trekkers and foragers

The major problem with attempts to typologize *terra firme* groups on the basis of their relative botanical knowledge is that such an exercise tends to disregard social, religious, and political considerations. Mobility is analysed as being caused either by environmental limitations (for cultural ecologists) or by historical constraints (for the proponents of cultural regression). In either case, ecological differences such as those between *terra firme* and *varzea* become historical differences, and mobility is seen as imposed from without. My main dissatisfaction with this model, as I have argued elsewhere (Rival 2002), is that it leaves no place for socio-cultural processes. Mobility is as much a product of historical will and religious belief as it is a form of adaptation to the environment or to historical circumstances. What deserves analytical attention is the fact that people decide to leave a resource-rich area for one which is relatively poorer in order to remain independent, to preserve a separate identity, or to resist assimilation, as in the Huaorani case. Whereas it cannot be denied that the Conquest favoured dispersion and fragmentation, the reasons for centrifugal processes are in part endogenous. It is through their decision-making and choice-exercising that social groups face historical forces, and, for that matter, environmental constraints as well. Mobility and the social forms it engenders need to be envisaged as part of the historical development of a distinct mode of life.

Mobility strategies need not be automatically related to foraging behaviour. Mobility is often used politically, as a way to segregate and avoid conflict. Many Amazonian anthropologists have noted a positive correlation between mobility and warfare, and between peace, gardening, and village life. Such correlation is commonly expressed not only as a set of contrastive practices, but also as a form of social discourse. Journet (1995), for example, notes that the Curripaco, who identify horticulture with peace and the foundation of society, equate the nomadic lifestyle of the Makú, seen as antithetical to culture and anterior to civilization, with warfare, hunting, and isolation in the forest. Fausto's (2001) study of two Parakanã groups who have chosen, after splitting, to live according to two divergent ways of life – nomadism and sedentism – illustrates the same association between pacific village life, horticulture, and sedentism, on the one hand, and mobility, foraging, warfare, and nomadism, on the other. Several ethnographers of the Yanomami point to the same close relation between intense warfare, a lack of internal differentiation, nomadism, and a lesser reliance on garden crops (Albert 1985; Colchester 1984; Ferguson 1995; Good 1989). As the archaeologist Susan Kent (1989) has shown, mobility is not automatically determined by subsistence strategies, even where groups with very different levels of socio-political organization and subsistence strategy co-exist.

Kent's observations (see also Kent & Vierich 1989: 124-30) are readily applicable to the Amazon context, and her generalizations about mobility go a long way to illuminate Huaorani trekking in terms of their own social understanding of space and of their relation to their forested environment, in particular the way in which mobility through the landscape relates to the use of old fallows. The Huaorani people of Amazonian Ecuador are best described as foragers. They practise a range of subsistence activities, including hunting, fishing, collecting wild fruits, tubers, and palm nuts, and exploit many other forest products such as honey and insects. They are characterized by a close-knit egalitarian social organization based on strong ties and shared communal patterns. Until ten years ago, and unlike their indigenous neighbours (all cultivating groups with a strong sense of identity as horticulturalists), the Huaorani chose to cultivate manioc and plantain only sporadically, and mainly for the preparation of ceremonial drinks. Even today, many Huaorani families prefer to secure their daily subsistence through hunting and gathering.

Through cycles of residential mobility, foraging activities, and the daily consumption of significant quantities of forest resources, the Huaorani are active agents in the concentration of useful species. Through marginal modifications (for instance, the quantity of seeds that they leave behind on abandoning a residential camp), they have enriched their habitat for humans, as well as for a number of non-human species. As such, they have contributed to the formation of anthropogenic forests of a very different kind from those created by swidden horticulturalists. The Huaorani are very conscious of past human activity, and are perfectly aware of the fact that every aspect of their forested territory has been transformed in equal measure by their ancestors, other indigenous groups, and the forces of nature and the supernatural. Taking the forest to be a legacy from the past, they have developed an understanding of the forest as owing its existence to past human activities. The forest exists to the extent that humans in the past lived and worked in it, and by so doing produced it as it is today for the benefit and use of the living. Their relation with the forest is lived as a social relation with themselves across generations, hence its eminently historical character. Trekking for the Huaorani is not simply a mundane activity relating to the pragmatics of subsistence and to environmental or historical adaptation, but, rather, a fundamental way of reproducing society through time (Rival 2002).

The Huaorani do not experience or represent their lack of intensive horticulture as a regression to a pre-social state (see Fausto 2001 for similar findings among the Parakanã). Their lesser reliance on garden products may well result from specific representations of the world and from political choices predating the Conquest. They contrast trekking and village life as two different types of sociality, linked to different styles of feasting and celebrating natural abundance. Moreover, they practise and represent trekking not only as a conscious form of adaptation to a landscape modified by past occupants, but also as a form of protection against predators. Animal predation and plant fructification thus become the contrastive sources of cultural representations which mediate the relation between environmental change and historical events. Huaorani trekking, the constant oscillation between a more horticultural and a more foraging mode of life, represents an unstable socio-political resolution half-way between two unsustainable extremes: peace and expansion; and war and destruction. Cultural creativity and political agency are therefore key components in the formation of anthropogenic forests (Rival 2002). Trekking, far from representing a necessary intermediary stage in the regression from sedentary agriculture to foraging, may, in some cases, constitute a *sui generis* solution to deep, contradictory forces of a political, religious, and social nature. Such internal processes may have long predated the Conquest and the disruptions it caused. I contend that a good understanding of Huaorani history and of its relationship with the natural history of the forest requires not only the examination of historiographic documents, as defended by post-colonial historians, and a botanical study of forest dynamics, as so cogently argued by Balée, but also an analysis of beliefs and values. For it is with such ideas in their minds that the Huaorani have become ecological and historical agents of change.

Balée has taken maximization theorists and other cultural ecologists to task for understanding adaptation to the environment in purely evolutionary terms, and for failing to recognize that a part of the Amazon forest to which indigenous people have adapted is not pristine, wild, or natural but, rather, an environment historically transformed by human production and consumption activities. While Balée's model illuminates some aspects of Amazonian historical ecology, the notion of cultural loss fails to explain the cultural logic embedded within a unique system of resource use, which drastically reduces dependence on cultivated crops. Like the Brazilian groups discussed by Balée, the Huaorani, for instance, have chosen to flee away from coercive powers and have adopted a wandering way of life. Although we will probably never know whether they descend from sedentarized and sophisticated cultivators, there is little doubt that they have wandered for centuries, and have always depended more on foraging than on agriculture. Their way of life and their long co-existence with more powerful, stratified, and complex agricultural societies cannot be explained with a model which aims to account for the progressive loss of cultivation by marginal tribes affected by disease, depopulation, and warfare. The reliance on resources created in the past by trekkers such as the Huaorani may be a characteristic shared by other Upper Amazon trekkers and foragers, such as the Makú, Nukak, and Hoti (Rival 1999). Highly mobile, nomadic, Amazonian societies have depended chiefly on hunting and gathering for centuries. People may cultivate, but they spend more time, and are far more interested in, collecting food from the forest than planting and cultivating. This cultural orientation and way of life cannot be explained away with reference to the environment and its conditionalities, or to history as a source of disruption and disintegration. A proper analysis of nomad foragers entails taking into consideration the anthropomorphic nature of their environment, as well as their cultural orientation, which strongly emphasizes life in the present.

Writing the history of landscapes

Research on the Amazon region has focused, perhaps more than in any other region of the world, on the history of the landscape. This is due to the fact that Amazonia is characterized, as Clement so rightly puts it, by 'a strong relationship between landscape and plant domestication' (1999: 191).[7] However, it is important to note the centrality of the concept of landscape for historical ecology as a whole. Landscape is the physical manifestation of the long-term human history of the environment. The goal is to document and understand the long-term creation of the environment as we know it today. This understanding may provide models for strategies of conservation and management of the environment in the present and future (Sutton & Anderson 2004: 27). Hardesty and Fowler write that historical ecology aims to describe 'landscape histories (ecohistories) at various scales and from various perspectives. This means searching the landscape for historical evidence of how they were created and changed by both human activity and natural processes, and the interactions between the two' (2001: 78). Crumley (2001: viii) defines landscape as the material manifestation of the relation between humans and the environment, and historical ecology, or landscape history, as the study of past ecosystems by charting the change in landscape over time. Writing the history of a particular landscape allows us to develop a form of anthropological inquiry that is more attentive to historical detail, as well as a kind of history which contains more holistic cultural and environmental data than normally used (see also Cronon 1995). As defined by historical ecologists, 'landscape' becomes a bridge between a whole range of disciplines and sub-disciplines pertaining to both the natural sciences and the humanities.

Sutton and Anderson (2004: 27) attribute historical ecology's special concern with landscape to cultural geographer Carl Sauer, who wrote the seminal 'The morphology of landscape' in 1925. For Sauer, landscape served as a common ground especially between anthropology, archaeology, and geography. It is important to remember the close collaboration that existed between Kroeber and Sauer in Berkeley during the 1930s and the 1940s. In her biography of Julian Steward, Kern mentions that Kroeber and Sauer

> offered seminars together, co-ordinated their research projects, and served as members of doctoral committees in each other's department. The two men shared many interests, particularly in cultural history, as well as a strongly Germanic worldview. ... At one time they discussed the possibility of a joint department, and Loeb, billed an 'anthropogeographer,' taught in both departments on occasion. ... For years, some graduate students from

anthropology considered geography a second home, and those from geography routinely took courses with Kroeber and Lowie ... (2003: 88-9).

Steward, who worked in both archaeology and ethnology during most of his professional life, was greatly influenced by both Kroeber and Sauer, and the recent trilogy on the cultivated landscapes of native America written by Carl Sauer's former students Denevan (2001), Doolittle (2000), and Whitmore and Turner (2001) combines insights from archaeology, physical and cultural geography, ethnography, and ecology. As Tsing remarks, studying nature in the making forces us to ask the question: 'How does human action shape nature, and how does nature shape human action?' (2001: 5). The concept of landscape enables historical ecologists to answer this question not only retrospectively, that is, with reference to 'the historically specific form of human and non human interaction that makes up the environmental project' (Tsing 2001: 5), but also with reference to the social, cultural, and political choices that are made today and that will influence the shape and the structure of future landscapes.

NOTES

[1] As defined here, Amazonia includes the lowland tropical and subtropical regions east of the Andes, that is, the Amazon Basin, the Orinoco Basin, and the Guiana shield.

[2] Other authors such as Gary Nabhan and Sara St Antoine (1993) speak of 'the extinction of experience,' and the link between such extinction and the processes of acculturation that promote language shift and cultural assimilation.

[3] 'Where the land is arable, horticulture is to be expected. The absence of horticulture requires an explanation that [need] not be based on environmental determinism' (Balée 1999: 26-7).

[4] 'The smaller a society gets, the more nomadic it becomes' (Balée 1992: 51).

[5] In the 'wild yam' controversy, Balée's thesis would therefore side with Bailey et al. (1989) and Bailey and Headland (1991) against Bahuchet, McKey, and de Garine (1991), for it supports the contention that no hunter-gatherer could have adapted to tropical rain-forest habitats without being surrounded by cultivators.

[6] '[P]robably all these forms of agriculture and agroforestry were present in the *terra firme* in a mosaic of variable population densities that may have included sectors of sparse semi-nomadic foragers; small but permanently settled households and extended families; and in some selected places large and permanent fields and associated villages ...' (Denevan 1996: 159-61).

[7] Although working with a very different kinds of data, Philippe Descola (1994) reaches a similar conclusion when he talks about the symbolic domestication of nature by the Achuar.

REFERENCES

ALBERT, B. 1985. Temps du sang, temps des cendres: représentation de la maladie, système rituel et espace politique chez les Yanomami du Sud-Est (Amazonie Brésilienne). Unpublished doctoral dissertation (Thèse de doctorat), Paris X: Nanterre.

BAHUCHET, S., D. McKEY & I. DE GARINE 1991. Wild yams revisited: is independence from agriculture possible for rainforest hunter-gatherers? *Human Ecology* **19**, 213-43.

BAILEY, R., G. HEAD, M. JENIKE, B. OWEN, R. RECHTMAN & E. ZECHENTER 1989. Hunting and gathering in a tropical rain forest: is it possible? *American Anthropologist* **91**, 59-82.
——— & T. HEADLAND 1991. The tropical rainforest: is it a productive environment for human foragers? *Human Ecology* **19**, 261-85.
BALÉE, W. 1988. Indigenous adaptation to Amazonian palm forests. *Principes* **32**: 2, 47-54.
——— 1989. The culture of Amazonian forests. In *Resource management in Amazonia: indigenous and folk strategies* (eds) D. Posey & W. Balée, 1-21. *Advances in Economic Botany* **7**. (Special issue).
——— 1992. People of the fallow: a historical ecology of foraging in Lowland South America. In *Conservation of neotropical forests – working from traditional resource use* (eds) K. Redford & C. Padoch, 35-57. New York: Columbia University Press.
——— 1993. Indigenous transformation of Amazonian forests: an example from Maranhão, Brazil. In *La remontée de l'Amazone* (eds) A.-C. Taylor & P. Descola. *L'Homme* **33**: 2-4, 235-58.
——— 1995. Historical ecology of Amazonia. In *Indigenous peoples and the future of Amazonia: an ecological anthropology of an endangered world* (ed.) L. Sponsel, 97-110. Tucson: University of Arizona Press.
——— 1998. Historical ecology. In *Advances in historical ecology* (ed.) W. Balée, 13-29. New York: Columbia University Press.
——— 1999. Sirionó. In *The Cambridge encyclopedia of hunters and gatherers* (eds) R. Lee & R. Daly, 77-85. Cambridge: University Press.
CLEMENT, C. 1999. 1492 and the loss of Amazonian crop genetic resources. I (The relation between domestication and human population decline) and II (Crop biogeography at contact). *Economic Botany* **53**, 177-216.
COLCHESTER, M. 1984. Rethinking stone age economics: some speculations concerning the pre-Columbian Yanomama economy. *Human Ecology* **12**, 291-314.
CRONON, W. 1995. The trouble with wilderness; or, getting back to the wrong nature. In *Uncommon ground: rethinking the human place in nature* (ed.) W. Cronon, 69-90. New York: W.W. Norton & Co.
CRUMLEY, C. 1998. Foreword. In *Advances in historical ecology* (ed.) W. Balée, ix-xiv. New-York: Columbia University Press.
——— (ed.). 2001. Introduction. In *New directions in anthropology and environment* (ed.) C. Crumley, vii-xi. Walnut Creek, Calif.: AltaMira Press.
DENEVAN, W. 1996. A bluff model of riverine settlement in prehistoric Amazonia. *Annals of the Association of American Geographers* **86**, 654-81.
——— 2001. *Cultivated landscapes of native Amazonia and the Andes*. New York: Oxford University Press.
DESCOLA, P. 1994. *In the society of nature: a native ecology in Amazonia*. Cambridge: University Press.
DOOLITTLE, W.F. 2000. *Cultivated landscapes of native North America*. New York: Oxford University Press.
FAUSTO, C. 2001. *Inimigos fiéis: história, guerra e xamanismo na Amazônia*. São Paulo: EDUSP.
FERGUSON, B. 1995. *Yanomami warfare: a political history*. Santa Fe: School of American Research.
GOOD, K. 1989. Foraging and farming among the Yanomami: can you have one without the other? Typescript.
HARDESTY, D.L. & D.D. FOWLER 2001. Archaeology and environmental changes. In *New directions in anthropology and environment* (ed.) C. Crumley, 72-89. Walnut Creek, Calif.: AltaMira Press.
HARRIS, D.R. 1996. Domesticatory relationships of people, plants and animals. In *Redefining nature: ecology, culture and domestication* (eds) R. Ellen & K. Fukui, 437-63. Oxford: Berg.
——— & G.C. HILLMAN (eds) 1989. *Foraging and farming: the evolution of plant exploitation*. London: Unwin Hyman.

HILL, J. & F. SANTOS-GRANERO 2002. Introduction. In *Comparative Arawakan histories: rethinking language family and culture area in Amazonia* (eds) J. Hill & F. Santos-Granero, 1-22. Urbana: University of Illinois Press.

HILL, K. & A.M. HURTADO 1996. *Aché life history: the ecology and demography of a foraging people*. New York: Aldine de Gruyter.

JOURNET, N. 1995. *La paix des jardins: structures sociales des indiens Curripaco du Haut Rio Negro, Colombie*. Paris: Institut d'Ethnologie, Musée de l'Homme.

KENT, S. 1989. Cross-cultural perceptions of farmers as hunters and the value of meat. In *Farmers and hunters: the implications of sedentism* (ed.) S. Kent, 1-17. Cambridge: University Press.

——— & H. VIERICH 1989. The myth of ecological determinism; anticipated mobility and site spatial organization. In *Farmers and hunters: the implications of sedentism* (ed.) S. Kent, 96-130. Cambridge: University Press.

KERN, V. 2003. *Scenes from the high desert: Julian Steward's life and theory*. Urbana: University of Illinois Press.

LATHRAP, D. 1970. *The upper Amazon*. London: Thames & Hudson.

——— 1973. The antiquity and importance of long-distance trade relationships in the moist tropics of pre-Columbian South America. *World Archaeology* 5, 170-86.

LEACH, E. 1954. *Political systems of highland Burma: a study of Kachin social structure*. London: Athlone Press.

LEHMAN, J., D.C. KERN, B. GLASER & W. WOODS (eds) 2003. *Amazonian dark earths: origins, properties, management*. Dordrecht: Kluwer Academic Publishers.

LÉVI-STRAUSS, C. 1948. La vie familiale et sociale des indiens Nambikwara. *Journal de la Société des Américanistes de Paris* 37, 1-132.

——— 1955. *Tristes tropiques*. Paris: Plon.

——— 1963a. The concept of archaism in anthropology. In *Structural anthropology I*, C. Lévi-Strauss, 101-19. New York: Basic Books.

——— 1963b. Do dual organizations exist? In *Structural anthropology I*, C. Lévi-Strauss, 132-67. New York: Basic Books.

MAFFI, L. 2001. Linking language and the environment: a co-evolutionary perspective. In *New directions in anthropology and environment* (ed.) C. Crumley, 24-48. Walnut Creek, Calif.: AltaMira Press.

MEGGERS, B. 1996 [1971]. *Amazonia: man and culture in a counterfeit paradise*. (Revised edition). Washington, D.C.: Smithsonian Institution Press.

NABHAN, G.P. & S. ST ANTOINE 1993. The loss of floral and faunal story: the extinction of experience. In *The biophilia hypothesis* (eds) S.R. Kellert & E.O. Wilson, 229-50. Washington, D.C.: Island Press.

POSEY, D. 1984. A preliminary report on diversified management of tropical rainforest by the Kayapó Indians of the Brazilian Amazon. *Advances in Economic Botany* 1, 112-26.

——— & W. BALÉE (eds) 1989. Resource managment in Amazonia: indigenous and folk strategies. *Advances in Economic Botany* 7. (Special issue).

RINDOS, D. 1984. *The origins of agriculture: an evolutionary perspective*. New York: Academic Press.

RIVAL, L. 1999. Introductory essay on South American hunters-and-gatherers. In *The Cambridge encyclopedia of hunters and gatherers* (eds) R. Lee & R. Daly, 77-85. Cambridge: University Press.

——— 2002. *Trekking through history: the Huaorani of Amazonian Ecuador*. New York: Columbia University Press.

ROOSEVELT, A. 1991. *Moundbuilders of the Amazon: geophysical archaeology on Marajó Island, Brazil*. San Diego: Academic Press.

——— 1993. The rise and fall of the Amazon chiefdoms. In *La remontée de l'Amazone* (eds) A.-C. Taylor & P. Descola. *L'Homme* 33: 2-4, 255-83.

——— 1994. Amazonian anthropology: strategy for a new synthesis. In *Amazonian Indians from prehistory to the present* (ed.) A. Roosevelt, 1-18. Tucson: University of Arizona Press.
——— 1998. Ancient and modern hunter-gatherers of lowland South America: an evolutionary problem. In *Advances in historical ecology* (ed.) W. Balée, 190-212. New York: Columbia University Press.
SAPIR, E. 1912. Language and the environment. *American Anthropologist* **14**, 226-42.
SAUER, C. 1925. The morphology of landscape. *University of California Publications in Geography* **2**: 2, 19-53.
STEWARD, J. 1946. Introduction. In *Handbook of South American Indians* (ed.) J. Steward, 1-15. (Bureau of American Ethnology, Bulletin **143**). Washington, D.C.: Government Printing Office.
SUTTON, M. & E.N. ANDERSON 2004. *Introduction to cultural ecology.* Oxford: Berg.
TSING, A.L. 2001. Nature in the making. In *New directions in anthropology and environment* (ed.) C. Crumley, 3-23. Walnut Creek, Calif.: AltaMira Press.
WHITMORE, T. & B.L. TURNER. 2001. *Cultivated landscapes of Middle America on the eve of Conquest.* New York: Oxford University Press.
YEN, D. 1989. The domestication of the environment. In *Foraging and farming: the evolution of plant exploitation* (eds) D.R. Harris & G. Hillman, 55-75. London: Unwin Hyman.

6

The interface between medical anthropology and medical ethnobiology

ANNA WALDSTEIN & CAMERON ADAMS

Medical anthropology is the study of the causes and consequences of sickness[1] in human beings, and its diverse theoretical orientations can be grouped into four major approaches: the medical ecology/biocultural approach, the political economy/critical medical anthropology approach, the interpretative or postmodern approach, and the ethnomedical approach. Medical ecology encompasses the contributions of biological anthropology to the understanding of human disease. An ecological perspective is employed to analyse the interaction of populations, pathogens, and environments at the core of disease processes (Armelagos, Leatherman, Ryan & Sibley 1992: 36; McElroy & Townsend 1996: 8). In contrast, critical medical anthropology is concerned primarily with political and economic factors in health and disease (Baer 1996: 452; Singer, Valentín, Baer & Jian 1992: 78-9), while interpretative medical anthropology focuses on metaphorical conceptions of the body and shows the social, political, and individual uses to which these conceptions are applied (Lock & Scheper-Hughes 1990: 50). Finally, ethnomedicine, following Fabrega (1975: 969), is the study of how members of different populations think about disease in cultural terms and organize themselves toward medical treatment. While all theoretical approaches to medical anthropology have addressed the causes of sickness, the study of how sickness is treated is largely confined to ethnomedicine.

An early focus for anthropologists concerned with sickness was on witchcraft and illness-causing spirits. Folk healers have been called excellent psychologists, and it has been argued that their therapy is so widely regarded as successful because psychosocial issues are addressed (Ackerknecht 1942: 514; Kleinman & Sung 1979: 24). Thus, the meanings of sickness events become a focal issue (see Good 1977) and are interpreted as structural microcosms of the society as a whole (Swedlund & Armelagos 1990).

For example, the Ndembu make a distinction between sicknesses caused by antagonistic entities and those arising from the natural progress of life. Maladies of the latter sort may be treated by herbalists, without a curative rite. However, the herbs used are said to have bitter, 'hot', or evil-smelling properties that disgust the sickness and drive it away (V. Turner 1967: 343). This led Turner to conclude that Ndembu curative procedures were primarily symbolic. Ethnomedical studies of shamanism, spirits, and religious curing rituals (which may mention the use of medicinal plants, but only in passing) are also common. Fabrega and Silver (1973), Levy (1983), Metzger and Williams (1970), and Edith Turner (1996) are prime examples of such studies. Alternatively, Elois Ann Berlin and Brent Berlin (1996) may be seen as a response to the heavy emphasis previously placed on the magical aspects of highland Maya ethnomedicine. The Berlins focus solely on the use of medicinal plants in the Chiapas highlands. Likewise, Green (1999) proposes that despite the way it is portrayed in the anthropological literature, African medicine is much more than witchcraft and sorcery, and supports his argument by examining naturalistic theories of contagious disease.

In this chapter we consider the role that medical ethnobiology has played in this shift of focus. We begin with a brief history of medical anthropology to illuminate why the empirical bases of ethnomedical systems were neglected for so long. We then review exemplary research in ethnophysiology and medical ethnobotany that contributes to the study of empirical aspects of medical systems, and conclude with suggestions for future research at the interface between medical ethnobiology and medical anthropology that we argue is likely to strengthen and advance both fields.

A history of the ethnomedical approach in medical anthropology

In possibly the first review article of the field, Scotch (1963: 31) does not define medical anthropology at all. Instead, he writes that medical anthropologists are people working in medical settings, or on problems of health and illness. Since the early 1960s medical anthropology has been defined and redefined many times. However, most medical anthropologists would probably agree that the field includes all formal anthropological research concerned with health, illness, and disease. In addition to reviewing the medical anthropology literature, Scotch (1963: 59) also suggested the establishment of a medical anthropology section in the American Anthropological Association.

However, medical anthropology has its roots in early ethnographies that included sections on the medical beliefs of non-Western cultures. While ethnographers such as Evans-Pritchard (1937; 1956) and Rivers (1924) were interested in medical activities, it was Ackerknecht (1942: 503) who first recognized that so-called 'primitive medicine' is not an immature or degenerate variety of biomedicine (or biomolecular medicine).[2] Nevertheless, he cautions that primitive empiricism is a myth and equates primitive medicine with magic (and hence irrational) medicine. According to Ackerknecht (1946: 478), medical behaviours that seem rational are really just instinctual or habitual.

In contrast, while British medical anthropologists were also preoccupied with theories of magical causality, they proposed that magical medicine is rational in the sense that models of diagnosis and treatment follow directly from theories of disease causation. This led to more emphasis on psychological, as opposed to physiological, efficacy of magical medicine and reduced ethnomedicine to the study of sorcery, magic, and ritual practices (Fortes 1976: xiii). To remedy this, the 1972 conference of the Association of Social Anthropologists was dedicated to ethnomedicine and participants were asked not to focus on witchcraft, sorcery, or magic. The volume that came out of the conference (Loudon 1976) was an early step toward filling a great void in the anthropological literature by looking at the mundane aspects of medicine in various parts of the world.

In 1976, when George Foster first made the distinction between *personalistic* (e.g. supernatural, wilful) and *naturalistic* (empirical) aspects of ethnomedical systems, the coexistence of empirical and magical models in medical systems was further acknowledged and elucidated. Personalistic aetiologies are based on the idea that the volition or intervention of an extra-natural force causes misfortune, including illness. The treatment of personalistic illnesses is the speciality of the shaman or spirit-medium, who conducts healing ceremonies aimed at appeasing angered gods or spirits, or counteracting the influence of witches or other shamans, and so on. Naturalistic causes are those sources of ill health that are the product of natural events or properties of natural substances, such as micro-organisms or an imbalance of hot and cold humours in the body. Treatments of conditions resulting from naturalistic causes are likely to be pragmatic and empirical. They usually involve medicinal preparations of plant or animal substances, prescribed by shamans, herbalists, physicians, and/or patients themselves. Most medical systems contain a mixture of personalistic and naturalistic strategies, and often a combination of strategies will be used in treating illnesses of either aetiology (Cosminsky 1977). Buckley (1985), Chavunduka (1994), Ortiz de Montellano (1990), and Snow (1993) are good examples of ethnomedical monographs which address both personalistic and naturalistic aspects of their respective medical systems.

In a related move, Kleinman (1980: 49-60) proposed that the internal structures of health care systems are roughly the same cross-culturally. He defines health care systems as local cultural systems composed of three overlapping parts: the popular, folk, and professional sectors. The popular sector involves self-treatment, or medicines given to a child by a parent. The folk sector involves folk/traditional healers (e.g. shamans). The professional sector includes specialists trained in formal medical schools. Biomedicine is a professional system, but it is not the only one. Ayurveda in India and Chinese medicine in China both have medical schools, are sponsored by the government, and are as well respected (if not more so) as biomedicine in these countries (see Hsu 1999; Leslie 1976). As in any model, these categories are fluid and all three can be present in a single medical system. However, medical anthropologists may decide to focus on only one of the three sectors. While this makes ethnomedical research more manageable, failure to consider differences between lay and specialist understandings of health and healing can lead to erroneous generalizations about medical systems (Tedlock 1987: 1072).

As witchcraft and illness-causing spirits stand out from other aspects of healing and are the most deviant from modern biomedicine, they are commonly the most attractive to anthropologists. Thus, much of the ethnomedical literature focuses on the personalistic aspects of the folk sector and on folk healers in particular. Indeed, most personalistic illnesses can be treated only by folk healers, because they alone know the proper rituals (see, e.g., Arkovitz & Manley 1990). In order to wrangle with these alien concepts, sickness is perceived by researchers as a meaning-rich event, and meaning appears to be of prime importance, to the exclusion of mundane bodily processes (Good 1977). It is difficult to approach illness from this perspective without being biased towards psychosomatic and psychological diseases (Fabrega 1974: 40; Rubel & Hass 1990: 119).

For example, early studies of highland Maya medicine very much follow the general trends of medical anthropology in that the focus is overwhelmingly biased towards the personalistic. From the earliest studies, the claim is made that most illnesses are thought to arise from witchcraft (e.g. Villa Rojas 1947; 1963). Shortly thereafter, a focus on healers leads to an examination of cosmology and the nature of the human being (see Cámara Barbachano 1966; Guiteras Holmes 1961; Vogt 1969; 1976). Because these studies are concerned with healers, rituals, and symbolic issues, the common illnesses of everyday life are ignored. Thus, personalistic ills and cures are the sole topic of discussion, and witchcraft and *nagualism* – belief in spirit helpers – take centre stage as the causes of illness. Finally, Harman and Kurtz (1982: 39) claim that naturalistic disease aetiology in highland Chiapas is an artefact of Protestant mission activity beginning in the 1940s.

In contrast to those approaches which emphasize illness and its treatment as a conceptual system, the ecological approach to disease tries to lay bare the environmental aspects of health. The ecological focus ultimately views pathogens and behaviour through an adaptive or evolutionary lens (see Alland 1970). Therefore, most attention is paid to naturalistic diseases or symptom clusters. However, the epidemiological model relied on by those in this camp tends to ignore social factors in disease events and its followers have set their gaze on only a limited slice of the ecosystem (Armelagos, Leatherman, Ryan & Sibley 1992: 38; Singer 1989: 226). Further, only illnesses labelled as such by the biomedical system tend to be studied, at the expense of illnesses not recognized in the West (Armelagos, Leatherman, Ryan & Sibley 1992: 40; Scheper-Hughes 1990: 192).

Prior to the recognition that there are two distinct categories of disease aetiology, researchers working on different aspects of sickness could only argue past one another. The descriptions of medical systems of non-Western populations are overly doused in rhetoric on the social meaning of illness, while pathogens are investigated regardless of what the host culture thinks about them. The impression that a non-ordained reader would come away with is that indigenous people are busy fending off spirits and warlocks while leaving themselves open to the ravages of the diseases afflicting them. Or, stranger still, it appears that everyone except those of European descent suffers only psychological or psychosomatic illnesses and the only needed cure is placebo. The major effect of Foster's and Kleinman's classification systems is that they allow researchers to define clearly what it is that they are investigating while not having to deny the presence of other domains of health. Owing to this opening of lines of communication, the shortcomings of both of the previous trends are laid bare.

In the real world, the medical cultural practices of particular populations may range from almost exclusively naturalistic, such as Western biomedicine, to the almost completely personalistic, as in the case of the Dobu (Fortune 1932), who consider even the smallest illness event to be caused by sorcery. However, it should be made clear that most populations appear to have a healthy mix of the two aetiologies. In many cultures, healing begins in the popular sector, especially if the patient is suffering from a minor, naturalistic affliction. Logan (1983) found that in the city of Juarez, Mexico, self-medication is the most frequently cited initial response to four common illness symptoms (headache, stomachache, cough, and diarrhoea) and to the folk illness *susto* (see below). However, healing in the popular sector has been largely ignored, with a notable exception being Elois Ann Berlin and Brent Berlin (1996). Additionally, there is usually more than one type of folk healer in a medical system (Huber 1990; Metzger & Willams 1963), and some folk healers use only naturalistic treatments. For example, in highland Bolivia, Qollahuaya folk healers are regarded as experts on (naturalistic) herbal medicine (Bastien 1982).

Like Foster's description of two disease aetiologies, Kleinman and Eisenberg further revolutionized the study of ethnomedical systems by clarifying the difference between illness and disease, the former defined as cultural facets of the latter bodily processes (Eisenberg 1977: 11; Kleinman 1973: 209; 1980: 72). Behaviours regarding disease can be seen as deriving from a society's ideas about illness (Sommerfeld 1994; Young 1986). It is in this body of work that we find the origins of biocultural medical anthropology, a methodology that attempts to integrate the complex interactions of beliefs and pathogens (see, e.g., Armelagos, Leatherman, Ryan & Sibley 1992; P.J. Brown & Inhorn 1990; Goodman & Leatherman 1998; Hahn & Kleinman 1983; Leatherman & Goodman 1997; McElroy 1990).

The strength of this perspective lies in combining the emic perspective of ethnomedicine with the etic measures of biomedical science. As noted by Browner, Ortiz de Montellano, and Rubel (1988: 682), some medical anthropologists have charged that it is not appropriate to apply biomolecular standards to ethnomedical systems, because to do so would be reductionist. However, there are practical benefits of using biomedical criteria to evaluate traditional medicine. For example, by conducting observational studies and blind, controlled clinical trials, researchers have been able to demonstrate that chiropractors are very effective in the treatment of certain spinal disorders. This led Anderson (1991: 2) to the conclusion that the lower status of chiropractors in relation to biomedical physicians reflects an artificial caste dynamic. That is, chiropractors are not inferior because they practise an ineffective form of healing. Likewise, the use of biomedical criteria can generate valuable new interpretations for comparative studies of human physiological processes, the ways in which such processes are perceived, and the culture-specific behaviours that these perceptions produce (Browner, Ortiz de Montellano & Rubel 1988: 683).

In fact, Browner, Ortiz de Montellano and Rubel (1988) developed a model for analysing ethnomedical data independently, as well as according to the standards of biomedicine. The first step is to identify the phenomena under investigation in emic terms. In the second step, one determines the extent to which the phenomena described can be understood in terms of biomedical concepts and methods. Finally, the third step is to identify the areas of convergence and divergence between the emically described phenomena and their biomedical understandings. At this stage biomedical concepts are used not to examine the phenomena in their own terms (as in step 2), but to see if they are consistent with biomedical assumptions. Another alternative to the reductionism of biomedical standards was used by García, Sierra, and Balám (1996) to study Yucatec Maya medicine. As two of these authors had experience with traditional Chinese medicine, they began to notice that many of the health concepts and practices they encountered among the Maya were understandable in terms of traditional Chinese medical concepts. Using a Chinese medical model allowed García and his colleagues to observe and study parts of traditional Maya medicine that might be overlooked or misunderstood by biomedicine.

An early success of this perspective was debunking the myth of culture-bound syndromes: illnesses thought to be unique to particular populations and regularly assumed to be psychological in nature (Hahn 1995; Simmons 1985). A prime example of this achievement is the discovery that *susto* – a commonly cited Mexican culture-bound syndrome – correlated with identifiable symptom clusters and high levels of stress (Rubel 1964; Rubel, O'Nell & Collado-Ardon 1984). In other words, *susto* is a culture-bound name for very real pathologies. Furthermore, Elois Ann Berlin and Jara (1993) and Luber (2002) address the folk illnesses *me' winik* ('mother of man') and *cha'lam tsots* ('second hair'), respectively. These ailments have been commonly considered to be culture-bound syndromes in highland Chiapas. However, *me' winik* is found to correlate strongly with biliary disease and *cha'lam tsots* with malnutrition.

Yet while medical anthropologists like Foster and Kleinman made it possible to define and study the mundane, empirical aspects of ethnomedical systems, they were not solely responsible for making this area of research an important and intellectually stimulating line of anthropological inquiry. The field of ethnobiology, particularly ethnophysiology and medical ethnobotany, has also made significant contributions. We propose that ethnophysiology illuminates the theoretical bases of ethnomedical systems by linking non-Western ideas of bodily processes with illness beliefs, while medical ethnobotany further confirms the empirical nature of these systems by showing that the use of plants as medicines is rational and systematic.

Ethnophysiology

Anthropological interest in the human body may best be summed up in the idea posed by Mary Douglas (1970: 93) that we have two bodies: the physical and the social. By this is meant that the body exists as a physical entity and may be considered as such, but also that the body, its parts, and their f unctions can be and are taken into the symbolic realm. Researchers generally take the body as a subject of study to be either one or the other of these conceptualizations.

Studies of the physical body are the domain of biological or physical anthropologists and generally focus on human evolution or the biological differences between populations, including ecological adaptation. This is the realm of the rules of Bergmann and Allen relating to limb and torso shape, of skin, eye, and hair morphology, brain development, and altitude effects on lung capacity. Unlike the study of the body as a physical entity, the approach to the social body is hardly unified. This stems from a multiplicity of scales and of social forms that can be considered. Because of its disparity, here we will focus only briefly on some of the more important trends. Perhaps the oldest and most studied aspect of the social body concerns adornment. To exemplify this, the Human Relations Area Files (n.d.) contains 9,296 citations regarding body alteration (code 304) and toilet (code 302). This does not include clothing, which is a separate and very rich category. Further, the ethnographic concern with the human body appears in the earliest ethnographies and is present to this day. An early departure from simple description of the body and how it is adorned is the work of Marcel Mauss (1973 [1935]). He posits the concept of 'techniques of the body', which are cultural aspects of bodily movement and posture. Bourdieu (1977) further refined this idea with his concept of *habitus*; the collection of learned bodily behaviours that are transferred from one generation to the next and which define a culture's physical patterns and the limits of innovation. Lyon and Barbalet (1994) and Martin (1994), for example, are more recent, and stronger, studies of the social construction of the body and its processes.

Linguistic anthropologists and students of ethnoscience were amongst the first to cross the intellectual divide between the physical and social bodies with several works on anatomical terminology and ethnoanatomy (see Petruck 1986; Stross 1976; Swanson & Witkowski 1977). Following the work of Brent Berlin and associates (B. Berlin 1974; 1976; B. Berlin, Breedlove & Raven 1968; 1973; 1974), Cecil H. Brown (1976) suggests twelve principles of body-part classification with four nomenclatural growth stages. Further, Cecil Brown and Stanley Witkowski (1981) suggest figurative universals to body-part nomenclature. For example, higher than chance frequencies for specific figurative labels, such as 'child of the eye' for pupil, are found in unrelated languages. This observation of linguistic universals for body parts suggests a strong link between the body itself and the cultural manifestation of body-part names.

Furthermore, body parts and the environment interact. Terms for environmental features are named with body parts (head of the river, foot of the mountain, etc.). This has been noted for the Mixe, who speak of soil as the flesh of the earth, rocks as its bones, and rivers as its veins and blood (Monaghan 1995). Similarly, as shown in the examples leading this paragraph, the same is true for English (Porteous 1986). However, the converse is also true: body parts are named for environmental features. For example, Bastien (1985) argues that the term for the bend of the knee among the Qollahuaya is derived from the term for a concave depression in the side of a mountain. Nevertheless, the former is far more common than the latter and supports further evidence of the anthropocentric character of human cognition of environmental phenomena (Ellen 2006 [1977]).

Meanwhile, medical anthropologists were studying the Latin American 'humoral', or hot/cold, system (reviewed in Foster 1976; 1984; 1994), which also focused on holistic conceptions of bodily process. Throughout Latin America much medical discourse is conducted in terms of hot or cold illnesses and their hot or cold cures, the underlying objective being to maintain a balance between the hot and cold poles. Though the actual system is more complex, the general rule is that cold remedies are used to treat hot illnesses, and vice versa. However, under increased scrutiny, the classification of remedies as either hot or cold proves to be inconsistent (Matthews 1983). Instead of a finely tuned adaptation to physical stress, the humoral system is more of a philosophical mnemonic to underscore empirical observations of plant efficacy (Foster 1994; Matthews 1983).

Ethnophysiology as a research topic has developed from these two diverse realms of inquiry. The linguistic work made it obvious that members of different cultures maintained inventories of body parts that were different from those of biomedicine. The work on the Latin American humoral system described what was clearly a theoretical basis of an indigenous medical practice. Together, they confirmed the observation that the way in which different body parts are believed to interact is related to the theoretical foundations of different medical cultures.

However, many medical anthropologists had claimed or assumed otherwise. An example from Chiapas is illustrative. The first mention of highland Maya concepts of the body is in the early work of Holland (1962; 1989 [1963]; Holland & Tharp 1964). One healer's rudimentary concepts of anatomy and physiology are described and it is concluded that 'the Tzotzil have only vague and elementary knowledge of the human body. ... The structure and function of bodily organs are poorly understood. The material components of the human body are [thought to be] simply flesh and bone' (Holland & Tharp 1964: 44). Shortly thereafter, Nash (1967: 133) argued that physiology plays no role in a curer's initial diagnosis of disease. This claim is echoed by Fabrega, Metzger, and Williams, who claim that 'the knowledge that underlies disease diagnoses and cure does not appear to depend in any significant way on an articulated set of native ideas about the structure and function of the body' (1970: 621). Fabrega and Silver (1973: 211) add that healers do not think about illness in terms of physiology, that the way in which a body is built and functions is not important to the medical system, and that Mayan traditional healers do not have specialized knowledge about the body.

However, descriptions of ethnophysiological systems continued to be reported and commonly displayed themes emanating from the broader world. For example, Villa Rojas (1980) argues that Yucatecan Maya conceive of the body as a microcosm of the universe. The organs are distributed in four quadrants around a central organ, called *tipte'*. This is a reflection of the universe, which is conceived as the four cardinal directions and a central point located at one's village. All of the organs are connected to the *tipte'*, which is further described as regulating all organ function. Again this reflects the belief that the religious behaviour of the community maintains the proper functioning of the universe.

More commonly, physiological process is described as reflecting earthly natural processes. Bastien (1985), for example, develops a model of physiology for the Qollahuaya of the Andes based on the topography and hydrologic processes of the mountain environment. As the Qollahuaya divide the mountain into three layers, based on altitude, so do they divide the body. Further, the process of water accumulating in caves and exiting the mountain from springs is a model for the belief that air, fat, water, and blood flow to the heart and are then redistributed throughout the body. As the body reflects the mountain, so the mountain reflects the body, with the top known as its head and the bottom as its foot. In Jamaica, the development of the body is perceived in terms of produce (Sobo 1993). With special emphasis on women, the life cycle of youth through puberty and sexual maturity to old age is described in terms of the development of fruit. Youth is seen as fruit that is not yet ripe. This is a stage believed to be dry, hard, and unformed. Puberty and sexual maturity are likened to ripe fruit that is plump, juicy, and soft yet firm. Finally, old age is like over-ripe fruit that is wrinkled, flaccid, and soft. These life stages of fruit are models from which ideals of body shape are derived. They also help a woman accept different phases of her body as part of a natural rhythm. Buckley's (1985) study of Yoruba medicine develops a three-colour symbolic scheme: black, white, and red. These colours are symbolically related respectively to soil, the sky, and clay, and, by extension, to skin, semen, and blood. The model for fertility is explained by the agricultural model of the sky (via rain) penetrating the topsoil and fertilizing the seeds in the clay. Symbolically, this is interpreted as white penetrating black and mixing with red to create life. Likewise, semen entering the black skin and mixing with the red blood inside produces offspring. Further, it explains why sex during menstruation is considered un-fertile, if not taboo. In this scenario white mixes with red as it seeps out from black. If the clay is turned out over the soil, the land is no longer fertile. Similarly, the Mixe consider the care of their fields to be equivalent to the care of a pregnant wife (Monaghan 1995). Planting of seed and watering the field are considered to be like introducing semen into a fertile womb. As such, the produce of the field is also, symbolically, one's offspring. Therefore, a gift of tortillas is a weighty gift indeed. Of course, the symbolic landscape is reflected upon the woman's body. If she were to get too wet or too dry, too hot or too cold, the foetus is thought to suffer the same effects that a corn plant would under similar circumstances. Agricultural analogies for conception are also common in India (Nichter & Nichter 1987: 19).

All these examples demonstrate that ethnophysiological models drive medical belief and behaviour. Further, all of them, with the exception of the first, deal exclusively with the naturalistic realm of health. As stated earlier, this realm has only recently become the focus of medical anthropologists. To illustrate the importance of this shift in focus, we return to the highland Maya example. Thus, Maffi notes the 'fundamental agreement in the literature that the body is not what healers centrally focus on and that closely heeding bodily symptoms in diagnosis may even detract from their reputation as experienced practitioners' (1994: 23). Traditional healers, it would seem, have a vested interest in ignoring or downplaying physiological symptoms when they profit from dealing with personalistic causes of illness such as witchcraft and soul loss. On the other hand, support for the argument that the general population does have a broad vocabulary for body parts is provided by Stross (1976), who further notes that there may be in excess of 100 terms for emotions that include 'heart' as part of the term. This suggests that part of the heart's role is in producing internal states such as emotions; a step towards collecting an ethnophysiology. Elois Ann Berlin and Enrique Castro (1988) make the case for a high level of detail in highland Maya knowledge of body parts, a knowledge that extends into the body as well as representing the surface. Laughlin (1988) has examined body-based metaphors, while Adams (2004) has described a highland Maya ethnophysiology at length, derived exclusivly from interviews with members of the general public, as opposed to specialists.

Medical ethnobotany
While research in ethnophysiology has exposed the local theoretical underpinnings of ethnomedical systems, medical ethnobotany[3] has advanced the study of naturalistic medicine by demonstrating that the use of medicinal plants in traditional societies is rational and systematic. Although much research in medical ethnobotany has been undertaken with the goal of identifying 'new' compounds for pharmaceutical development (Leonti, Ramirez, Sticher & Heinrich 2003: 219), the documentation and evaluation of medicinal plants has also been used to demonstrate that indigenous healers and other people who use medicinal plants actually employ species that are biologically effective. Moreover, other areas of ethnobotanical research, including human chemical ecology, zoopharmacognosy, medicinal plant selection criteria, and

universal patterns in the use of plants as medicines, directly address anthropological questions and portray indigenous peoples as active observers of, and experimenters with, their natural environments. Examples of research in these areas follow.

Documentation and evaluation of medicinal plants used in ethnomedical systems

Though ethnobotanical inventories often present botanical (and sometimes pharmacological) data abstracted from their cultural contexts, and contain little analysis or interpretation (Ellen 1996: 457; Etkin 1988a: 24), they are a testament to the elaborate character of indigenous pharmacopoeias. Most inventories are also excellent sources of raw data that can be used to answer theoretical questions and conduct cross-cultural studies of medicinal plant use. We might consider a few good examples of the many that are now available. Bandoni, Mendiondo, Rondina, and Coussio (1972) surveyed fifty-six Argentine plants used as popular medicines, presenting data on common names, habitats, medicinal uses, and chemical compositions. Chinemana, Drummond, Mavi, and de Zoysa (1985) conducted household surveys in Zimbabwe and documented over fifty plant species used in self-treatment, which previously had been ignored by researchers of Zimbabwean traditional medicine. Basualdo, Zardini, and Ortiz (1995) identified underground storage organs of seventeen plant species that are sold as medicines in Paraguay, used to treat a variety of health conditions and as refreshing beverages. Inventories of medicinal flora have also been made in the East Black Sea region of Turkey (Fujita et al. 1995), the Myagdi District of Nepal (Manandhar 1995), three regions of Zaïre (the Democratic Republic of the Congo) (Disengomoka & Delaveau 1983; Disengomoka, Delaveau & Sengele 1983) and Lucca Province in central Italy (Pieroni 2000). More recently, Hernández Cano and Volpato (2004) investigated traditional herbal mixtures in Eastern Cuba and present data on 170 species used in the preparation of 199 composite medicines.

Other authors have published excellent medical ethnobotanical inventory monographs. Boulos (1983) has attempted to bridge the gap between folk medicine and modern medicine with reference to the medicinal plants of North Africa. He collected plants in Egypt, Libya, Tunisia, Morocco, and Algeria from 1952 to 1980. His volume includes 369 indigenous, naturalized, and/or cultivated species, alphabetically arranged according to ninety-seven families. For each plant the distribution in North Africa, scientific name, vernacular name (in Arabic, Berber, English, and French when available) and uses (including part used and sometimes preparation method) are given. Moerman (1986) has compiled and organized data from numerous ethnobotanical works on Native American medicinal plants. For each of the species included, information is presented on the cultural group and indication, as well as notes on how it is used/prepared. Hutchings (1996) has compiled data on Zulu medicine and medical ethnobotany in order to stimulate research that will validate traditional claims made about medicinal plants and facilitate the acceptance of traditional healing in South Africa.

Some ethnobotanical inventories go even further and include information on the bioactivity of medicinal plants. For example, Schultes and Raffauf (1990) have inventoried 1,516 medicinal and toxic plants used by peoples of the northwest Amazon. For each species, the scientific name, distribution and references are given, as well as local names, local uses, chemical compounds, and pharmacological activities, as available. Caceres, Cano, Samayoa, and Aguilar (1990) screened eighty-four plants commonly used to treat gastrointestinal conditions in Guatemala against five pathogenic enterobacteria. Their article includes both ethnobotanical data (scientific name, local name[s] and parts used) and results of the enterobacteria screens for each of these species. Caceres and his colleagues conclude that their results confirm the basis of folk uses of these plants for gastrointestinal disease. Finally, in an ethnographically rich study of Garífuna ethnobotany in Nicaragua, Coe and Anderson (1996) have documented 229 species in the Garífuna pharmacopoeia. They discuss the chemical constituents of these plants and include a long table listing species, common names, uses, medicinal applications, part used, mode of preparation, mode of administration, and results of alkaloid tests.

While standard pharmacological evaluations of medicinal plants often support their use as traditional medicines, some of the key researchers at the interface between medical ethnobotany and anthropology have suggested that the way efficacy is measured must be culturally appropriate. Efficacy has been defined as something that 'works' by directly or indirectly producing a set of required, culturally defined, outcomes (Etkin 1988*b*: 301). Etkin (1988*b*) suggests that the key to any consideration of efficacy is the distinction between its emic and etic interpretations. The efficacy of medicinal plants may be judged on their ability to induce full remission of symptoms. However, in many cultures other physical signs, such as fever, salivation, or emesis (i.e. proximate outcomes), are also important, as they indicate that the plant has initiated the healing process. Thus, even when medicinal plants with biologically active constituents are the focus of study, the failure of biomedical researchers to consider the cultural context within which plants are used can result in misunderstandings of a plant's efficacy.

Bernard Ortiz de Montellano (1982) was one of the first anthropological ethnobotanists to suggest that traditional medicines must be evaluated according to the standards of folk medical systems, rather than those of biomedicine. He evaluated whether chemical components in twenty-five Aztec medicinal plants could produce the effects ascribed to them by Aztec physicians. He found that sixteen would produce effects claimed in native sources, four may possibly be active, and five do not appear to possess activity. This finding led Ortiz de Montellano to conclude that although magic and religion were important in Aztec medicine, there was also a strong empirical underpinning.

In a later study, Ortiz de Montellano and Browner (1985) developed a method for assessing the efficacy of medicinal plants according to both indigenous understandings of their therapeutic effects and the standards of biomedicine. The first step in their method requires investigating informants' own understandings of illness aetiology to predict what medicines would be considered appropriate for their treatment. These ethnomedical data are then combined with data on chemical constituents, physiological effects, and biomedical concepts that constitute four 'confidence levels' for the efficacy of plant medicines. Level I comprises reported folk use. Multiple reports of use by populations widely dispersed through space, or persistent reports over long periods of time, increase the probability that a plant will exhibit pharmacological activity. Level II plants meet the criteria of level I and show the desired activity of isolated compounds or extracts in *in vitro* or *in vivo* tests. At level III, plants satisfy level II requirements and show a plausible biochemical mechanism by which the active constituents could exert the indicated physiological effect. Finally, level IV plants fulfil the criteria for level III and have been clinically tested, or are commonly used in medicine. Information on a plant's level of confidence is then considered in the light of its emic evaluation of efficacy. Ortiz de Montellano and Browner demonstrate the use of this model in their evaluation of plants used for postpartum recovery, uterine bleeding, infertility, and dysmenorrhoea in highland Oaxaca. They conclude that their informants use plants that contain bioactive chemicals, which would allow them to achieve the desired effects.

Human chemical ecology, zoopharmacognosy and the evolutionary basis of medicinal plant use

Chemical ecology is the study of ecological factors that influence the chemical composition of plants and the coevolutionary relationships between plant and animal species. Chemical ecology brings an ecological and evolutionary perspective to the study of medicinal plant use. It has been suggested that the coevolution of human-plant units is an appropriate theoretical framework for understanding the interrelationships between plant secondary compounds and humans in both the past and the present (Jackson 1991: 511). Johns (1990) uses a chemical ecological model to elucidate the ways in which humans select plants for consumption on the basis of chemical constituents (i.e. chemical selection). He begins by proposing a model for human chemical ecology which includes the idea that: (1) humans seek to maintain physiological homeostasis through maximizing the beneficial effects of ingested components and minimizing the effects of potential toxins; (2) natural selection has produced the interrelated physiological and behavioural mechanisms that allow humans to deal with environmental chemicals; (3) humans have unique cultural traits that play a role in their interactions with plant chemical constituents (i.e. language and technological innovations, including plant domestication, are particularly powerful forces); and (4) the character of human interactions with plant chemicals occurs within, and is influenced by, a broader ecological framework. This model is used to consider: biological adaptations for dealing with plant toxins; technological methods of detoxification; domestication; human perception, cognition, and behaviour in relation to plant chemicals; and plant chemical defences as determinants of the human diet (Johns, Kokworo & Kimanani 1990).

Human food procurement is constrained by the same secondary products that make plants unavailable as food for other species. While humans could select for plants with fewer of these compounds through domestication, such plants would be vulnerable to insect attacks and disease. In contrast, food-processing technology eliminates undesirable compounds after a plant has matured, allowing chemicals to play their natural defensive role during the development of the plant (Johns & Kubo 1988). Johns and Kubo (1988) suggest that processing methods show many similarities worldwide and can be classified according to the way in which they function to eliminate toxins. Common processing techniques include detoxification by heating, solution, fermentation, adsorption, drying, and chemical reaction due to pH change. In another article, by Johns and Keen (1986), plant domestication is viewed as a good focus for the study of human chemical ecology. This is because domestication involves direct interactions between humans and specific compounds over time and its history is recorded in genetic changes brought about in plants. Johns and Keen test the hypothesis that Aymara cultivators have placed direct selection pressure on the chemical constituents of potatoes through their attention to taste. Specifically, the study focuses on glycoalkaloid perception. The authors report that their results support the idea that the Aymara are capable of making judgements of potato quality on the basis of taste. However, these judgements do not appear to be based (primarily) on glycoalkaloid content. Jackson (1991) has also reviewed the literature on metabolic changes in humans that are induced by chronic exposure to specific plant chemicals.

While much of the human chemical ecology literature is related to dietary practices, chemical ecology is also important for understanding medicinal plant use. This is because it provides a framework for linking compounds and substances produced by plants with changes in human physiology. Because secondary metabolites produced by one species influence the conditions for other species, they can also affect health conditions in humans. Thus, the principles of chemical ecology help explain traditional medicine (Torssell 1997: 43). Johns (1990) explores the chemical-ecological basis of medicine, including the interrelationship of food and medicine, the biological and cultural basis of human medicine, and the acquisition of empirical medical knowledge. To build his arguments for an evolutionary and ecological basis of ethnomedical systems, he cites the use of medicinal plant species by non-human primates. Research in 'zoopharmacognosy' reflects interactions between animals and secondary compounds. It shows that self-medication in plants is not unique to humans and has evolutionary depth.

One of the earliest reports of possible self-medication by chimpanzees was made by Wrangham and Nishida in 1983. While most plants are eaten quickly and continuously, chimpanzees eat *Aspilia* spp. leaves slowly, one at a time, and do not chew them. In Gombe National Park, these leaves were eaten early in the morning. Chimpanzees were observed eating *Aspilia* leaves in all months except April and November and in periods of both food abundance and food scarcity. *Aspilia* was eaten by individuals of both sexes and from different social groups. Wrangham and Nishida speculated that these leaves may be eaten in the morning because they have stimulant properties.

Leaf-swallowing and bitter pith-chewing are the most common chimpanzee self-medication behaviours (Huffman 1997; Huffman & Wrangham 1994). Newton and Nishida (1990) note that leaf-swallowing is very similar to buccal and sublingual administration techniques in human medicine. A buccal route of administration might avoid the degradation of active compounds in the stomach. Further, the diurnal pattern of *Aspilia* consumption, observed by Wrangham and Nishida (1983), may be due to diurnal variation in the concentration of pharmacological compounds, rather than the presence of stimulant properties (Newton & Nishida 1990). Indeed, while pharmacological analyses of *Aspilia* spp. have not found evidence of stimulants in this genus, thiarubrine A (a potent antibiotic compound) has been isolated from two species of *Aspilia* (Rodriguez et al. 1985: 420). Chimpanzee leaf-swallowing has also been correlated with the presence of tapeworm fragments in chimp dung. Chimpanzees with tapeworms tend to swallow leaves, which cause tapeworm fragments to be shed. However, this does not necessarily mean that leaf-swallowing affected the viability or reproductive success of tapeworms (Wrangham 1995). Currently, leaf-swallowing has been observed with over thirty-four different plant species at over thirteen great ape study sites. These plant species all have bristly, rough-surfaced leaves (Huffman 2001: 657).

Huffman and Seifu (1989) documented changes in the apparent health and behaviour of one chimpanzee before and after she sucked out and swallowed the bitter juice from the pith of another medicinal plant, *Vernonia amygdalina*. Humans in several African populations use *V. amygdalina* for malaria, schistosomiasis, amoebic dysentery, parasites, and stomachaches, and phytochemical analyses of this species show that it contains several sesquiterpene lactones and steroid glucosides. The leaves of *Vernonia* can be lethal if ingested raw in large amounts. Yet while domesticated goats have been known to poison themselves by eating this plant, chimpanzees avoid the leaves in favour of the bitter pith (Huffman 2001: 656).

Based on the current literature on chimpanzee self-medication and Johns's work on human chemical ecology, Huffman (2001) speculates on the evolution of traditional medicine in humans. Given the range of self-medicative behaviour in extant apes, early hominids likely displayed similar behaviours. Thus, the fundamentals of perceiving the medicinal properties of a plant by its organoleptic qualities have roots deep in our primate history. Language would have enabled people to share and pass on experiences of plant properties and their effects against disease. The adoption of food preparation and detoxification technologies would have reduced many secondary chemicals from the human diet and this could have contributed to a rise in some diseases that would have otherwise been kept in check. This may have led humans to use some plants specifically as medicines and/or medicinal foods. Increased disease and stress brought about by domestication, increased population density, and sedentary lifestyles could have been the stimulus for the elaboration of medical practices and botanical pharmacopoeias.

Universal patterns, medicinal plant selection criteria and the development of indigenous pharmacopoeias

Further insight into the elaboration of botanical pharmacopoeias comes from comparative studies on medicinal floras of various regions, as well as research on the criteria that local peoples use to select medicinal plants. Moerman, Pemberton, Kiefer, and Berlin (1999) compared the medicinal floras of Chiapas, Korea, Kashmir, North America, and Ecuador. The authors ranked medicinal families by their residuals, which are the differences between numbers of medicinal species in a family predicted by regression analysis and the true ethnographically determined number of medicinal species. They found that, with rare exception, the top five families of any of the four holarctic regions (Chiapas, North America, Kashmir, and Korea) rank in the top quarter of any of the others. However, there was significantly less overlap between the medicinal plants of neotropical Ecuador and the other four floras. The results suggest that as related peoples migrated from Asia into the Americas, they passed through related floras and selected similar species as medicinals. However, when people reached the tropical zone, the flora was different enough that they had to develop a new range of ethnobotanical knowledge. The implications of this research include the possibility that people speaking unrelated languages have generated similarly useful knowledge over broad geographic regions or that knowledge of medicinal plants which was carried from Asia to America during the upper Paleolithic has persisted throughout the past 15,000 years.

More recently, Leonti, Ramirez, Sticher, and Heinrich (2003) examined the pharmacopoeia of the Popoluca using the same method as Moerman, Pemberton, Kiefer, and Berlin (1999). The Popoluca inhabit the Sierra de Santa Marta, Veracruz, Mexico, an area with holarctic and neotropical floristic overlap. The authors used Moerman's method of regression analysis to identify the top and bottom five medicinal families. The top five medicinal families were Asteraceae, Piperaceae, Fabaceae, Euphorbiaceae, and Lamiaceae, while the five least medicinal families were Orchidaceae, Poaceae, Rubiaceae, Cyperaceae, and Moraceae. Diversity of secondary compounds, taste and smell properties, and morphology distinguish Asteraceae, Lamiaceae, and Fabaceae from Poaceae and Cyperaceae. Asteraceae and Lamiaceae are highly ranked in other holarctic pharmacopoeias, while Piperaceae and Fabaceae are highly ranked in the Ecuadorian neotropical pharmacopoeia. While the flora of Veracruz is most similar to that of Chiapas, it is more closely related to the neotropical flora of Ecuador than to any of the other three holarctic flora studied by Moerman, Pemberton, Kiefer, and Berlin (1999). However, the Popolucan pharmacopoeia is generally more closely related to holarctic pharmacopoeias. Leonti, Ramirez, Sticher, and Heinrich (2003) conclude that common medicinal plant knowledge may be due to common selection criteria, not just shared common knowledge.

The study of medicinal plant selection criteria goes back to some of the early work on the Latin American humoral (hot/cold) system. However, as described above, the classification of a plant as hot or cold appears to be variable. That is, illnesses are perceived to be either hot or cold and the plants that treat them are thus presumed to be the opposite. Thus, medicinal plants do not appear to be selected based on their humoral qualities. This observation has led researchers working in Latin America to consider other factors important to the selection of medicinal species. For example, in her study of the choice of herbal remedies for female reproductive conditions in a Chinantec township of Oaxaca, Browner (1985) reviews studies and hypotheses about humoral medicine in Latin America and suggests that other factors may influence the choice of treatment. She found that herbal medicines for nine reproduction-related ailments fell into two broad categories: (1) medicines for uterus expulsion, which work by opening the womb through warming, irritating, or drying qualities; and (2) medicines for uterine retention, which close the uterus by cooling, removing impurities from the blood, or healing a woman's womb. Browner then describes in detail various reproductive events (pregnancy, birth, postpartum recovery, and contraception) as perceived by her informants, along with the rationale for using various medicinal plants during these events. She concludes that the hot/cold theory does not adequately represent the ethnomedical system of her informants and (in the case of reproduction) falls within a more comprehensive framework of expulsion or retention.

Other researchers have considered the role of plants' organoleptic (taste, smell, morphology) characteristics on the selection of medicinal species. Brett (1994) examines the role of taste, odour, and irritation in chemoperception and medicinal plant selection in the Tzeltal municipality of Cancuc, Chiapas, Mexico. Taste is the most important sense involved in medicinal plant selection among the Tzeltal. For the Tzeltal, when one reports that a plant has a certain flavour, 'inherent in that concept is information about the illness categories for which the plant would serve as a useful medicine, which in turn informs on the likely preparation, administration and probably mode of action' (Brett 1994: 198).

Ankli, Sticher, and Heinrich (1999) investigated taste and smell perceptions of both medicinal and non-medicinal plants, as well as hot/cold classification among the Yucatec Maya. They worked with forty healers and midwives, including twelve key informants who considered themselves herbalists. They collected voucher specimens and information on uses, preparation, application, and properties of the plants and descriptions of illnesses. Healers were then asked to collect plants with no medicinal value, and information on the taste, smell, and humoral quality of these plants was recorded. Ankli and her colleagues found that non-medicinal plants were more often reported to have no smell or taste. In 50 per cent of cases a good odour was characteristic of medicinal plants and thus a sign of medicinal use. A larger percentage of medicinal plants were reported to be astringent or sweet, but both medicinal and non-medicinal plants were reported to be bitter. Again, humoral qualities appeared to refer not to selection criteria but to a plant's classification, and non-medicinal plants were rarely classified humorally. Ankli and her colleagues then attempted to correlate Mayan perceptions of taste and smell properties with known chemical consituents. Plants considered to be astringent contain polyphenols, whose disinfecting properties would be useful in the treatment of infections like diarrhoea and skin diseases. Bitter-tasting plants are very common, and bitterness may be attributed to a variety of compounds, including cardenolides, terpenes, sesquiterpene lactones, alkaloids, and saponins. All aromatic plants contain large amounts of essential oils. No particular group of constituents was found to be responsible for unpleasant smells in plants considered to be bad-smelling. Likewise, no specific group(s) of compounds was associated with alleged hot or cold properties of plants. Ankli and her colleagues concluded that taste and smell are important selection criteria for medicinal plants among the Maya, but they are not a central unifying principle of Maya medicinal plant classification.

A similar study of Popolucan criteria for distinguishing between medicinal and non-medicinal plants was undertaken by Leonti, Sticher, and Heinrich (2002). This team of researchers worked with eight healers and asked them (as a group) to identify plants with no curative properties and plants with curative properties. Informants were later interviewed individually and asked to identify smell and taste properties of the plants, their virtues, and healing concepts in general. These data were used to identify several cultural parameters that the Popoluca use to describe medicinal plants for specific illness categories. For example, most remedies for diarrhoea and dysentery are astringent. Bitter plants are generally used to treat skeleto-muscular pain and stomachache, but are considered to be somewhat risky due to possible overdose. Medicines for coughs are composed of sweet plants and are often sweetened with sugar or honey. Sour plants are associated with a cold humoral state and are used for hot conditions like fever, headache, and pain while urinating. It is interesting to note that a similar pattern in the plants/mixtures used for certain conditions are found in Cuba; mixtures for stomach ulcers and diarrhoea tending to have high tannin content (Hernández Cano & Volpato 2004: 309). Leonti, Sticher, and Heinrich (2002) also compared taste and smell properties in medicinal versus non-medicinal plants and found a significant difference. This finding suggests that the selection of medicinal plants is closely correlated with these species' organoleptic properties. Moreover, some healers considered non-medicinal species potentially medicinal if they had interesting taste or smell properties.

While research on the organoleptic qualities of plants and the selection of medicinal species is intriguing and provides further evidence that indigenous peoples actively experiment with medicinal plants, it is limited by the fact that nearly all of the published work in this area has been done in Mexico. Two notable exceptions are recent studies which look at chemosensory perception in Borneo (Gollin 2004) and Peru (Shepard 2004). Gollin (2004) notes that bitter and astringent plants have received the most attention in the study of organoleptic properties of medicinal plants, as they are widespread and indicate the presence of alkaloids and tannins. Her work brings our attention to local categories of sensation that are difficult to gloss in English. *Nglidah* is one such category among the Kenyah Leppo' Ke and seems to correlate with plants high in volatile and essential oils. This group of compounds includes camphor and menthol, which ease pain but are also understood by the Kenyah Leppo' Ke to open pores that allow internal wind and heat to escape. Shepard (2004) describes the role of taste, irritation, odour, and visual/tactile properties in medicinal plant selection in two culturally and linguistically distinct villages in the rainforest of southeastern Peru. He also shows how the theoretical frameworks of the ethnomedical systems in these villages are mediated by and reflected in sensation.

To summarize, medical ethnobotany has contributed to the study of naturalistic aspects of ethnomedical systems in a number of ways. Inventories of medicinal flora demonstrate that local peoples are familiar with large numbers of species in their environments. Evaluations of the efficacy of such plants generally validate their use as medicines, especially when they are judged by indigenous expectations and criteria. This line of research shows that the use of medicinal plants is much more than simply symbolic. Moreover, research in human chemical ecology and the self-medication behaviours of chimpanzees suggest that the use of medicinal plants has evolutionary roots and is as old as humankind itself. Over the course of this long relationship with medicinal plants, certain characteristics of these plants have become well known as producing particular effects. This mid-range theory of organoleptic signs and their healing properties becomes further elaborated into grand theories of health and its maintenance, including ethnophysiologies.

Conclusion

We can summarize our argument by returning to the highland Maya case one last time. While specialist healers do not focus on bodily symptoms of disease, Stross (1976) showed that the general highland Maya population of Chiapas has a broad vocabulary for body parts, and Elois Ann Berlin and Enrique Castro (1988) demonstrated that this includes internal organs. Meanwhile, Maffi's (1994) treatise on highland Maya ethnosymptomology makes an attempt to consider the points of intersection between Tzeltal disease concepts and those of biomedicine. Though some of the Tzeltal explanations and concepts miss the target according to biomedical wisdom, they still point to real and consistent clusters of symptoms that are meaningful and precise. The ethnosymptomology encodes particular health concerns and provides for particular and consistent responses to them. Adams (2004) draws on both these lines of inquiry in his study of highland Maya ethnophysiology and shows that Maya understanding of how the body functions underlies popular beliefs about health and illness. Moreover, research on medical ethnobotany in Chiapas (E.A. Berlin & Berlin 1996; E.A. Berlin et al. 1995; E.A. Berlin, Berlin, Tortoriello, Meckes & Villareal 1995; Brett 1994; Casagrande 2002; Stepp 2002; Stepp & Moerman 2001) further confirms the empirical character of highland Maya medicine. This literature shows that plant selection does not exhibit random patterns of distribution and that medicinal plants do display bioactivity for their claimed usage in clinical trials. Given these two bodies of research, highland Maya medicine is a system based on observation and experimentation that is theoretically driven, a far cry from the world of cosmology and witchcraft as described through the 1960s.

Finally, the heavy Americanist bias in this chapter is not simply a reflection of our country of origin. Although perceptions of physiological processes and their relations to theories of disease are documented in the medical texts of Old World professional medical systems, ethnophysiological accounts in relation to ethnomedical systems based on oral traditions are surprisingly infrequent. Likewise, while there are ethnobotanical inventories of medicinal plants from all over the world (and studies of chimpanzee self-medication from Africa), virtually all of the work on human chemical ecology and medicinal plant selection criteria has been done in the Americas. Thus, research on ethnophysiology, chemical ecology, and medicinal plant selection criteria in Asia, Africa, Europe, and Oceania is needed to determine whether all ethnomedical systems are empirical and theoretically driven. We also encourage anthropologists who find that their interests lie at the interface of medical anthropology and medical ethnobiology to pursue research that specifically links ethnophysiology with medical ethnobotany, as we are aware of only one published study (Quinlan, Quinlan & Nolan 2002) which explicitly does so. Such research will lead to a more sophisticated understanding of efficacy, medicine, and the body and contribute to broader questions in ethnobiology, especially those concerned with how local peoples perceive, understand, classify, and use resources in their environments.

NOTES

[1] Generally, in medical anthropology the term 'disease' refers to pathological states, whether or not they are culturally recognized, and are the arena of the biomedical model. An 'illness' is a person's perception and experiences of socially devalued states, including (but not limited to) disease (cf. Kleinman 1980). This dichotomy implies that non-biomedical systems do not address physiological manifestations of ill-health. Kleinman uses sickness (a less loaded term) as a blanket term to label events involving disease and/or illness.

[2] Here we will use the terms 'biomedicine', or 'biomolecular medicine' to refer to the dominant medical system in the industrial world, which is based on biological investigations of disease at the molecular level. This is not to imply that there is a dichotomy between biomedicine and ethnomedicine. Biomolecular medicine is a cultural construct as much as any other medical system.

[3] Many of the pioneers in the study of the naturalistic aspects of ethnomedicine – Elois Ann and Brent Berlin, Browner, Etkin, Ortiz de Montellano – might be considered both medical anthropologists and ethnobotanists.

REFERENCES

ACKERKNECHT, E.H. 1942. Problems of primitive medicine. *Bulletin of the History of Medicine* 11, 503-21.

――― 1946. Natural diseases and rational treatment in primitive medicine. *Bulletin of the History of Medicine* **19**, 467-97.
ADAMS, C. 2004. *The ethnophysiology of the Tzeltal Maya of highland Chiapas*. Doctoral dissertation, University of Georgia.
ALLAND, A. 1970. *Adaptation in cultural evolution: an approach to medical anthropology*. New York: Columbia University Press.
ANDERSON, R. 1991. The efficacy of ethnomedicine: research methods in trouble. *Medical Anthropology* **13**, 1-17.
ANKLI, A., O. STICHER & M. HEINRICH 1999. Yucatec Maya medicinal plants versus nonmedicinal plants: indigenous characterization and selection. *Human Ecology* **27**, 557-80.
ARKOVITZ, M.S. & M. MANLEY 1990. Specialization and referral among the N'anga (traditional healers) of Zimbabwe. *Tropical Doctor* **20**, 109-10.
ARMELAGOS, G.J., T. LEATHERMAN, M. RYAN & L. SIBLEY 1992. Biocultural synthesis in medical anthropology. *Medical Anthropology* **14**, 35-52.
BAER, H. 1996. Toward a political ecology of health in medical anthropology. *Medical Anthropology Quarterly* **10**, 451-4.
BANDONI, A.L., M.E. MENDIONDO, R.V. RONDINA & J.D. COUSSIO 1972. Survey of Argentine medicinal plants: I. folklore and phytochemical screening. *Lloydia* **35**, 69-80.
BASTIEN, J. 1982. Herbal curing by Qollahuaya Andeans. *Journal of Ethnopharmacology* **6**, 13-28.
――― 1985. Qollahuaya-Andean body concepts: a topographical-hydraulic model of physiology. *American Anthropologist* **87**, 595-611.
BASUALDO, I., E.M. ZARDINI & M. ORTIZ 1995. Medicinal plants of Paraguay: underground organs, II. *Economic Botany* **49**, 387-94.
BERLIN, B. 1974. Further notes on covert categories and folk taxonomies: a reply to Brown. *American Anthropologist* **76**, 327-31.
――― 1976. The concept of rank in ethnobiological classification: some evidence from Aguaruna folk botany. *American Ethnologist* **3**, 381-99.
―――, D.E. BREEDLOVE & P.H. RAVEN 1968. Covert categories and folk taxonomies. *American Anthropologist* **70**, 290-9.
―――, ――― & ――― 1973. General principles of classification and nomenclature in folk biology. *American Anthropologist* **75**, 214-42.
―――, ――― & ――― 1974. *Principles of Tzeltal plant classification*. New York: Academic Press.
BERLIN, E.A. & B. BERLIN 1996. *Medical ethnobiology of the highland Maya of Chiapas, Mexico: the gastrointestinal diseases*. Princeton: University Press.
―――, ―――, M. MECKES, J. TORTORIELLO, M.-L. VILLAREAL & X. LOYOLA 1995. The scientific basis of diagnosis and treatment of gastrointestinal diseases by the Tzeltal and Tzotzil Maya of Southern Mexico. In *Naked science* (ed.) L. Nader, 43-68. London: Routledge.
―――, ―――, J. TORTORIELLO, M. MECKES & M.-L. VILLAREAL 1995. Spasmolytic activity of medicinal plants used to treat gastrointestinal and respiratory diseases in the highlands of Chiapas. *Phytomedicine* **2**, 57-66.
――― & E. CASTRO 1988. *Etnoanatomía Tzeltal-Tzotzil, Chiapas, Mexico*. San Cristóbal de las Casas, Chiapas: PROCOMITH.
――― & V. JARA 1993. Me' winik: discovery of the biomedical equivalence for a Maya ethnomedical syndrome. *Social Science & Medicine* **37**, 671-8.
BOULOS, L. 1983. *Medicinal plants of North Africa*. Algonac, Mich.: Reference Publications, Inc.
BOURDIEU, P. 1977. *Outline of a theory of practice*. (Cambridge Studies in Social Anthropology **16**). Cambridge: University Press.
BRETT, J.A. 1994. *Medicinal plant selection criteria among the Tzeltal Maya of highland Chiapas, Mexico*. Ph.D. dissertation, University of California at San Francisco.

BROWN, C.H. 1976. General principles of human anatomical partonomy and speculations on the growth of partonomic nomenclature. *American Ethnologist* 3, 400-24.

——— & S.R. WITKOWSKI 1981. Figurative language in a universalist perspective. *American Ethnologist* 8, 596-615.

BROWN, P.J. & M.C. INHORN 1990. Disease, ecology, and human behavior. In *Medical anthropology: a handbook of theory and method* (eds) T.M. Johnson & C.F. Sargent, 187-214. New York: Praeger.

BROWNER, C. 1985. Criteria for selecting herbal remedies. *Ethnology* 24, 13-32.

———, B.R. ORTIZ DE MONTELLANO & A.J. RUBEL 1988. A methodology for cross-cultural ethnomedical research. *Current Anthropology* 29, 681-9.

BUCKLEY, A.D. 1985. *Yoruba medicine*. Oxford: Clarendon Press.

CACERES, A., O. CANO, B. SAMAYOA & L. AGUILAR 1990. Plants used in Guatemala for the treatment of gastrointestinal disorders: 1. screening of 84 plants against enterobacteria. *Journal of Ethnopharmacology* 30, 55-73.

CÁMARA BARBACHANO, F. 1966. *Persistencia y cambio cultural entre Tzeltales de los altos de Chiapas: estudio comarativo de las Instituciones religiosas y políticas de los municipios de Tenejapa y Oxchuc*. México, D.F.: Escuela Nacional de Antropología e Historia.

CASAGRANDE, D.G. 2002. Ecology, cognition, and cultural transmission of Tzeltal Maya medicinal plant knowledge. Ph.D. dissertation, University of Georgia.

CHAVUNDUKA, G. 1994. *Traditional medicine in modern Zimbabwe*. Harare: University of Zimbabwe Publications.

CHINEMANA, F., R.B. DRUMMOND, S. MAVI & I. DE ZOYSA 1985. Indigenous plant remedies in Zimbabwe. *Journal of Ethnopharmacology* 14, 159-72.

COE, F.G. & G.J. ANDERSON 1996. Ethnobotany of the Garífuna of eastern Nicaragua. *Economic Botany* 50, 71-107.

COSMINSKY, S. 1977. The impact of methods on the analysis of illness concepts in a Guatemalan community. *Social Science and Medicine* 11, 325-32.

DISENGOMOKA, I. & P. DELAVEAU 1983. Medicinal plants used for child's respiratory diseases in Zaïre: part I. *Journal of Ethnopharmacology* 8, 257-63.

———, ——— & K. SENGELE 1983. Medicinal plants used for child's respiratory diseases in Zaïre: part II. *Journal of Ethnopharmacology* 8, 265-77.

DOUGLAS, M. 1970. *Natural symbols: explorations in cosmology*. London: Barrie & Jenkins.

EISENBERG, L. 1977. Disease and illness: distinctions between professional and popular ideas of sickness. *Culture, Medicine, and Psychiatry* 1, 9-24.

ELLEN, R. 1996. Putting plants in their place: anthropological approaches to understanding the ethnobotanical knowledge of rainforest populations. In *Tropical rainforest research: current issues* (eds) D.S. Edwards, W.E. Booth & S.C. Choy, 457-65. Dordrecht: Kluwer Academic Publishers.

——— 2006 [1977]. Anatomical classification and the semiotics of the body. In *The categorical impulse: essays in the anthropology of classifying behaviour*, R.F. Ellen, 90-116. Oxford: Berghahn.

ETKIN, N.L. 1988a. Ethnopharmacology: biobehavioral approaches in the anthropological study of indigenous medicines. *Annual Review of Anthropology* 17, 23-42.

——— 1988b. Cultural constructions of efficacy. In *The context of medicines in developing countries: studies in pharmaceutical anthropology* (eds) S. VanDerGeest and S.R. Whyte, 299-327. Dordrecht: Kluwer Academic Publishers.

EVANS-PRITCHARD, E.E. 1937. *Witchcraft, oracles, and magic among the Azande*. Oxford: Clarendon Press.

——— 1956. *Nuer religion*. Oxford: Clarendon Press.

FABREGA, H. 1974. *Disease and social behavior: an interdisciplinary perspective*. Cambridge, Mass.: MIT Press.

——— 1975. The need for an ethnomedical science. *Science* 189, 969-75.

―――, D. METZGER & G. WILLIAMS 1970. Psychiatric implications of health and illness in a Maya Indian group: a preliminary statement. *Social Science and Medicine* **3**, 609-26.

―――― & D. SILVER 1973. *Illness and shamanistic curing in Zinacantan: an ethnomedical analysis.* Stanford: University Press.

FORTES, M. 1976. Foreword. In *Social anthropology and medicine* (ed.) J. Loudon, ix-xx. London: Academic Press.

FORTUNE, R. 1932. *The sorcerers of Dobu.* New York: Dutton.

FOSTER, G.M. 1976. Disease etiologies in non-western medical systems. *American Anthropologist* **78**, 773-82.

―――― 1984. The concept of 'neutral' in humoral medical systems. *Medical Anthropology* **8**, 181-94.

―――― 1994. *Hippocrate's Latin American legacy: humoral medicine in the New World.* (Theory and Practice in Medical Anthropology and International Health 1). Langhorne, Penn.: Gordon & Breach.

FUJITA, T., E. SEZIK, M. TABATA, E. YESILADA, G. HONDA, Y. TAKEDA, T. TANAKA & Y. TAKAISHI 1995. Traditional medicine in Turkey VII: folk medicine in middle and west Black Sea regions. *Economic Botany* **49**, 406-22.

GARCÍA, H., A. SIERRA & G. BALÁM 1996. *Medicina Maya tradicional: confrontación con el systems conceptual Chino.* Mexico: EDUCE.

GOLLIN, L. 2004. Subtle and profound sensory attributes of medicinal plants among the Kenyah Leppo' Ke of East Kalimantan, Borneo. *Journal of Ethnobiology* **24**, 173-201.

GOOD, B.J. 1977. The heart of what's the matter: the semantics of illness in Iran. *Culture Medicine and Psychiatry* **1**, 25-58.

GOODMAN, A. & T.J. LEATHERMAN 1998. Traversing the chasm between biology and culture: an introduction. In *Building a new biocultural synthesis* (eds) A. Goodman & T.J. Leatherman, 3-41. Ann Arbor: University of Michigan Press.

GREEN, E.C. 1999. *Indigenous theories of contagious disease.* Walnut Creek, Calif.: Altamira Press.

GUITERAS HOLMES, C. 1961. *Perils of the soul; the world view of a Tzotzil Indian.* New York: Free Press of Glencoe.

HAHN, R. 1995. *Sickness and healing: an anthropological perspective.* New Haven: Yale University Press.

―――― & A.M. KLEINMAN 1983. Biomedical practice and anthropological theory: frameworks and direction. *Annual Review of Anthropology* **12**, 305-33.

HARMAN, R.C. & W.B. KURTZ 1982. Effects of missionary and government programs on medical behavior: a Maya Indian study. In *Ancient and modern medical practices in Mesoamerica* (eds) B. Ortiz de Montellano, J.R. Moriarty, III, R.C. Harman, W.B. Kurtz & B.J. McMahan, 39-52. (KATUNOB. Occasional Publications in Mesoamerican Anthropology). Greely: University of Northern Colorado.

HERNÁNDEZ CANO, J. & G. VOLPATO 2004. Herbal mixtures in the traditional medicine of eastern Cuba. *Journal of Ethnopharmacology* **90**, 293-316.

HOLLAND, W.R. 1962. Highland Maya folk medicine: a study of culture change. Ph.D. dissertation, University of Arizona.

―――― 1989 [1963]. *Medecina Maya en los altos de Chiapas.* México, D.F.: Instituto Nacional Indigenista.

―――― & R.G. THARP 1964. Highland Maya psychotherapy. *American Anthropologist* **66**, 41-52.

HSU, E. 1999. *The transmission of Chinese medicine.* Cambridge: University Press.

HUBER, B.R. 1990. The recruitment of Nahua curers: role conflict and gender. *Ethnology* **29**, 159-76.

HUFFMAN, M.A. 1997. Current evidence for self-medication in primates: a multidisciplinary perspective. *Yearbook of Physical Anthropology* **40**, 171-200.

―――― 2001. Self-medicative behavior in the African great apes: an evolutionary perspective into the origins of human traditional medicine. *BioScience* **51**, 551-61.

―――― & M. SEIFU 1989. Observations on the illness and consumption of a possibly medicinal plant *Vernonia amygdalina* (Del.), by a wild chimpanzee in the Mahale Mountains National Park, Tanzania. *Primates* **30**, 51-63.

―――― & R.W. WRANGHAM 1994. Diversity of medicinal plant use by chimpanzees in the wild. In *Chimpanzee cultures* (eds) R.W. Wrangham, W.C. McGrew, F.B.M. deWaal & P. Heltne, 129-48. Cambridge, Mass.: Harvard University Press.

HUMAN RELATIONS AREA FILES n.d. Available on-line (*http://www.yale.edu/hraf*; accessed 15 February 2004).

HUTCHINGS, A. 1996. *Zulu medicinal plants: an inventory*. Pietermaritzburg: University of Natal Press.

JACKSON, F.L.C. 1991. Secondary compounds in plants (allelochemicals) as promotors of human biological variability. *Annual Review of Anthropology* **20**, 205-46.

JOHNS, T. 1990. *With bitter herbs they shall eat it: chemical ecology and the origins of human diet and medicine*. Tucson: University of Arizona Press.

―――― & S.L. KEEN 1986. Taste evaluation of potato glycoalkaloids by the Aymara: a case study in human chemical ecology. *Human Ecology* **14**, 437-52.

――――, J.O. KOKWORO & E.K. KIMANANI 1990. Herbal remedies of the Luo of Siaya District, Kenya: establishing quantitative criteria for consensus. *Economic Botany* **44**, 369-81.

―――― & I. KUBO 1988. A survey of traditional methods employed for the detoxification of plant foods. *Journal of Ethnobiology* **8**, 81-129.

KLEINMAN, A. 1973. Medicine's symbolic reality. *Inquiry* **16**, 206-13.

―――― 1980. *Patients and healers in the context of culture*. Berkeley: Univeristy of California Press.

―――― & L.H. SUNG 1979. Why do indigenous practitioners successfully heal? *Social Science and Medicine* **13B**, 7-26.

LAUGHLIN, R. (ed.) 1988. *The great Tzotzil dictionary of Santo Domingo Zinacantán*. (Smithsonian Contributions to Anthropology **31**). Washington, D.C.: Smithsonian Institution Press.

LEATHERMAN, T.L. & A. GOODMAN 1997. Expanding the biocultural synthesis toward a biology of poverty. *American Journal of Physical Anthropology* **102**, 1-3.

LEONTI, M., F. RAMIREZ, O. STICHER & M. HEINRICH 2003. Medicinal flora of the Popoluca, Mexico: a botanical systematical perspective. *Economic Botany* **57**, 218-30.

――――, O. STICHER & M. HEINRICH 2002. Medicinal plants of the Popoluca, México: organoleptic properties as indigenous selection criteria. *Journal of Ethnopharmacology* **81**, 307-15.

LESLIE, C. (ed.) 1976. *Asian medical systems*. Berkeley: University of California Press.

LEVY, J. 1983. Traditional Navajo health beliefs and practices. In *Disease, change and the role of medicine* (ed.) S. Kunitz, 118-45. Berkeley: University of California Press.

LOCK, M. & N. SCHEPER-HUGHES 1990. A critical-interpretive approach in medical anthropology: rituals and routines of discipline and dissent. In *Medical anthropology: contemporary theory and method* (eds) T.M. Johnson & C.F. Sargent, 47-72. New York: Praeger.

LOGAN, K. 1983. The role of pharmacists and over the counter medications in the health care system of a Mexican city. *Medical Anthropology* **7: 2**, 68-84.

LOUDON, J.B.H. 1976. Introduction. In *Social anthropology and medicine* (ed.) J.B.H. Loudon, 1-48. London: Academic Press.

LUBER, G.E. 2002. The biocultural epidemiology of 'second-hair' illness in two Mesoamerican societies. Ph.D. dissertation, University of Georgia.

Lyon, M.L. & J.M. Barbalet 1994. Society's body: emotion and the 'somatization' of social theory. In *Embodiment and experience: the existential ground of culture and self* (ed.) T.J. Csordas, 48-66. Cambridge: University Press.

McElroy, A. 1990. Biocultural models in studies of human health and adaptation. *Medical Anthropology Quarterly* 4, 243-65.

———— & P. Townsend 1996. *Medical anthropology in ecological perspective*. Boulder, Colo.: Westview.

Maffi, L. 1994. A linguistic analysis of Tzeltal Maya ethnosymptomatology. Ph.D. dissertation, University of California at Berkeley.

Manandhar, N.P. 1995. An inventory of some herbal drugs of Myagdi District, Nepal. *Economic Botany* 49, 371-9.

Martin, E. 1994. *Flexible bodies*. Boston: Beacon Press.

Matthews, H.F. 1983. Context-specific variation in humoral classification. *American Anthropologist* 85, 826-47.

Mauss, M. 1973 [1935]. Techniques of the body. *Economy and Society* 2, 70-88.

Metzger, D. & G. Willams 1963. Tenejapa medicine I: the role of the curer. *Southwestern Journal of Anthropology* 19, 216-34.

———— & ———— 1970. Medicina Tenejapaneca. In *Ensayos de antropologia en la zona central de Chiapas* (eds) N.A. McQuown & J. Pitt-Rivers, 391-441. México, D.F.: Instituto Nacional Indigenista.

Moerman, D.E. 1986. *Medicinal plants of native America*. Ann Arbor: University of Michigan Museum of Anthropology.

————, R. Pemberton, D. Kiefer & B. Berlin 1999. A comparative analysis of five medicinal floras. *Journal of Ethnobiology* 19, 49-67.

Monaghan, J. 1995. *The covenants with earth and rain: exchange, sacrifice, and revelation in Mixtec sociality*. Norman: University of Oklahoma Press.

Nash, J. 1967. The logic of behavior: curing in a Maya Indian town. *Human Organization* 26, 132-40.

Newton, P.N. & T. Nishida 1990. Possible buccal administration of herbal drugs by wild chimpanzees, *Pan troglodytes*. *Animal Behavior* 39, 798-801.

Nichter, M. & M. Nichter 1987. Cultural notions of fertility in South Asia and their impact on Sri Lankan family planning practices. *Human Organization* 46, 18-28.

Ortiz de Montellano, B. 1982. Empirical Aztec medicine: Aztec medicinal plants seem to be effective if they are judged by Aztec standards. In *Ancient and modern medical practices in Mesoamerica*. (eds) B. Ortiz de Montellano, J.R. Moriarty, III, R.C. Harman, W.B. Kurtz & B.J. McMahan, 18-22. (KATUNOB. Occasional Publications in Mesoamerican Anthropology). Greely: University of Northern Colorado.

———— 1990. *Aztec medicine, health, and nutrition*. New Brunswick: Rutgers University Press.

———— & C. Browner 1985. Chemical bases for medicinal plant use in Oaxaca, Mexico. *Journal of Ethnopharmacology* 13, 57-88.

Petruck, M.R.L. 1986. Body part terminology in Hebrew: a study in lexical semantics. Ph.D. dissertation, University of California, Berkeley.

Pieroni, A. 2000. Medicinal plants and food medicines in the folk traditions of the Upper Lucca Province, Italy. *Journal of Ethnopharmacology* 70, 235-73.

Porteous, J.D. 1986. Bodyscape: the body-landscape metaphor. *The Canadian Geographer* 30, 2-12.

Quinlan, M.B., R.J. Quinlan & J.M. Nolan 2002. Ethnophysiology and herbal treatments of intestinal worms in Dominica, West Indies. *Journal of Ethnopharmacology* 80, 75-83.

RIVERS, W.H.R. 1924. *Medicine, magic and religion.* London: Kegan Paul, Trench & Turner.
RODRIGUEZ, E., M. AREGULLIN, T. NISHIDA, S. UEHARA, R. WRANGHAM, Z. ABRAMOWSKI, A. FINLAYSON & G.N.H. TOWERS 1985. Thiarubrine A, a bioactive constituent of Aspilia (Asteraceae) consumed by wild chimpanzees. *Experientia* **41**, 419-20.
RUBEL, A.J. 1964. The epidemiology of a folk illness: *susto* in Hispanic America. *Ethnology* **3**, 268-83.
────── & M.R. HASS 1990. Ethnomedicine. In *Medical anthropology: a handbook of theory and method* (eds) T.M. Johnson & C.F. Sargent, 115-31. New York: Praeger.
──────, C.W. O'NELL & R.C. COLLADO-ARDON 1984. *Susto: a folk illness.* Berkeley: University of California Press.
SCHEPER-HUGHES, N. 1990. Three propositions for a critically applied medical anthropology. *Social Science and Medicine* **30**, 189-97.
SCHULTES, R.E. & R.F. RAFFAUF 1990. *The healing forest: medicinal and toxic plants of the Northwest Amazon.* Portland: Dioscorides Press.
SCOTCH, N. 1963. Medical anthropology. *Biennial Review of Anthropology* **3**, 30-68.
SHEPARD, G.H. 2004. A sensory ecology of medicinal plant therapy in two Amazonian societies. *American Anthropologist* **106**, 252-66.
SIMMONS, R.C. 1985. Sorting the culture-bound syndromes. In *The culture-bound syndromes* (eds) R.C. Simmons & C.C. Hughes, 25-40. Dordrecht: D. Reidel.
SINGER, M. 1989. The limitations of medical ecology: the concept of adaptation in the context of social stratification and social transformation. *Medical Anthropology* **10**, 223-34.
──────, F. VALENTÍN, H. BAER & Z. JIAN 1992. Why does Juan García have a drinking problem? The perspective of critical medical anthropology. *Medical Anthropology* **14**, 77-108.
SNOW, L.F. 1993. *Walkin' over medicine.* Boulder, Colo.: Westview Press.
SOBO, E.J. 1993. *One blood: the Jamaican body.* Albany: State University of New York Press.
SOMMERFELD, J. 1994. Emerging epidemic diseases. In *Disease in evolution: global changes and the emergence of infectious diseases* (eds) M.E. Wilson, R. Levins & A. Spielman, 276-84. (Annals of the New York Academy of Sciences). New York: New York Academy of Sciences.
STEPP, J.R. 2002. Highland Maya medical ethnobotany in ecological perspective. Ph.D. dissertation, University of Georgia.
────── & D.E. MOERMAN 2001. The importance of weeds in ethnopharmacology. *Journal of Ethnopharmacology* **71**, 19-23.
STROSS, B. 1976. Tzeltal anatomical terminology: semantic processes. In *Mayan linguistics* (ed.) M. McClaran, 243-67. Los Angeles: UCLA, American Indian Studies Center.
SWANSON, R.A. & S. WITKOWSKI 1977. Hopi ethnoanatomy: a comparative treatment. *Proceedings of the American Philosophical Society* **121**, 320-37.
SWEDLUND, A.C. & G.J. ARMELAGOS 1990. Introduction. In *Disease in populations in transition: anthropological and epidemiological perspectives* (eds) G.J. Armelagos & A.C. Swedlund, 1-24. New York: Bergin & Garvey.
TEDLOCK, B. 1987. An interpretive solution to the problem of humoral medicine in Latin America. *Social Science & Medicine* **24**, 1069-83.
TORSSELL, K. 1997. *Natural product chemistry: a mechanistic, biosynthetic, and ecological approach.* Stockholm: Apotekarsocieteten, Swedish Pharma.
TURNER, E. 1996. *The hands feel it: healing and spirit presence among a northern Alaskan people.* De Kalb: Northern Illinois University Press.
TURNER, V. 1967. *The forest of symbols: aspects of Ndembu ritual.* Ithaca, N.Y.: Cornell University Press.
VILLA ROJAS, A. 1947. Kinship and nagualism in a Tzeltal community, southeastern Mexico. *American Anthropologist* **49**, 578-87.

——— 1963. El nagualismo como recurso de control socialde los grupos mayances de Chiapas México. *Estudios de Cultura Maya* **3**, 243-60.

——— 1980. La imagen del cuerpo humano segun los Mayas de Yucatan. *Anales de Antropología* **XVII**, 31-46.

VOGT, E.Z. 1969. *Zinacantan: a Maya community in the highlands of Chiapas*. Cambridge, Mass.: Belknap Press of Harvard University Press.

——— 1976. *Tortillas for the gods: a symbolic analysis of Zinacanteco rituals*. Cambridge, Mass.: Harvard University Press.

WRANGHAM, R.W. 1995. Relationship of chimpanzee leaf-swallowing to a tapeworm infection. *American Journal of Primatology* **37**, 297-303.

——— & T. NISHIDA 1983. *Aspilia* spp. leaves: a puzzle in the feeding behavior of wild chimpanzees. *Primates* **24**, 276-82.

YOUNG, A. 1986. Internalizing and externalizing medical belief systems: an Ethiopian example. In *Concepts of health, illness and disease: a comparative perspective* (eds) C. Currer & M. Stacey, 137-61. Oxford: Berg.

7
Ethnobiology and applied anthropology: *rapprochement* of the academic with the practical

PAUL SILLITOE

Until recently ethnobiology and applied anthropology have not appeared obvious bedfellows. Ethnobiology has become for anthropologists a quintessentially academic pursuit interested largely in human cognitive capacities explored through exotic classifications of flora and fauna, whereas applied anthropology has sought a practical use for the discipline's knowledge, firstly in colonial contexts and subsequently, to a large extent, in the area of international development. These two fields illustrate the distinction between the 'pure' and the 'practical' that has dogged anthropology for decades (Foster 1969: 39-42, 131-44; Grillo 1985: 8-9; Keen 1999: 33-5), an unfortunate distinction that we may be in the process of exorcizing from the discipline. In this respect I agree with the sentiment expressed by Whitehead, that

> Science is a river with two sources, the practical source and the theoretical source. The practical source is the desire to direct our actions to achieve predetermined ends ... The theoretical source is the desire to understand. I most emphatically state that I do not consider one source as in any sense nobler than the other, or intrinsically more interesting. I cannot see why it is nobler to strive to understand than to busy oneself with the right ordering of one's actions. Both have bad sides; there are evil ends directing actions, and there are ignoble curiosities of the understanding (cited in Bennett 1976: x).

However, I also recognize that this acceptance that the 'pure' and the 'practical' may coexist represents a particular deep-rooted challenge for anthropology.

The bringing together of the applied and the academic has proved difficult for anthropology. A brief historical review of the discipline's attempts to find applications – as it was obliged to go through recurrent rounds of trying to show its usefulness and secure funding – indicates the intractability. Opportunities for the discipline to demonstrate its relevance were few. Either it was asked to do things that it was not able to undertake competently by those ignorant of the subject's limitations; or it was shut out from work that it might have contributed to significantly, such as international development, by institutional structures that discriminated against it. After repeated rebuffs and bad experiences trying to find an applied niche, anthropology largely retreated in the second half of the twentieth century to its academic redoubt.

It is possible that the current association of ethnobiology with applied anthropology is going to break with this series of rebuffs. The advent of participatory approaches to development, and particularly the emergence of so-called 'indigenous knowledge inquiries', presents us, I think, with a better opportunity to show our mettle than ever before. Ethnobiology has contributed significantly to the recognition of indigenous knowledge in development, a point of association with applied anthropology. It promises to lead to a clearer definition of the discipline's application, namely show its relevance. Unlike applied anthropology, ethnobiology is a clearly defined area of enquiry promoting interdisciplinary research, encompassing a range of other subjects with an interest in the utility of folk biological knowledge. These other subjects are challenging us to think in new ways about applying anthropology and leading in new directions. Whereas the 'old' applied anthropology was obliged to work in top-down contexts with a focus on social institutions, the 'new' applied anthropology, heavily influenced by persons working in the ethnobiological field, is bottom-up and concerns local knowledge systems and ethical issues about the ownership of intellectual property.

A brief history of the academic and practical aspects of ethnobiology

Ethnobiology and applied anthropology have considerable, though largely, until now, separate, histories. While the term 'ethnobiology' did not come into use until the twentieth century, naturalists have shown an interest in local biological knowledge since the time Europeans started to explore the world from the fifteenth century onwards. Indeed this knowledge informed biological science from early on, as witnessed in the research of Linnaeus, who corresponded with people around the world, and Rumphius' work in Southeast Asia (Ellen 2004). In a sense we can interpret this as an early example of applied anthropology linked with ethnobiology as Europeans not only sought to understand the new regions they intruded into but also were on the look-out for resources that they might profitably exploit, engaging in practices that today we should consider tantamount to biopiracy. Many new crops, for instance, entered into Europe during this period, such as the potato, tomato, pumpkin, maize, and tobacco. This tradition continued on through from the seventeenth to the twentieth centuries, as in the work of Darwin, who during the *Beagle* voyage took interest in local biological information, and Wallace, who showed a similar interest during his sojourn in the Malay archipelago (now Indonesia). Indeed these scholars comprised a strand of what eventually became the discipline of anthropology, some early ethnographers such as Haddon and Boas being natural scientists who became more interested in studying the local communities in which they found themselves than their natural environments.

Research in this tradition continues to this day, with ethnobiologists working in close association with local populations regularly turning up evidence of more species than previously known to science, such as Gary Martin and his collaborators (Martin, Agama, Beaman & Nais 2002: 12-13) on Mount Kinabalu in Sabah (northern Borneo), whose ethnobotanical project increased by 90 per cent, in just four years of collaborative research, the number of known palm taxa documented in the previous 145 years. Sometimes such work turns up new species to science, such as in the New Guinea highlands, where I collected a new species of mycrohylid frog (*Choerophryne* sp) still unfortunately not formally scientifically named and described because I could not find more than one, unlike Ralph Bulmer, who, on the other side of the central cordillera, has a hylid frog named after him, *Litoria bulmeri*, a fitting tribute to his path-breaking ethno-herpetological work. Those who engage in bio-prospecting, hoping, for example, to find elsewhere plants with unknown medicinal or cosmetic properties, are also current-day descendants of this tradition, although from a humanistic perspective such work gives rise to worries about theft of others' knowledge (see, e.g., contributions in Svarstad & Dhillion 2000).

Regarding applied anthropology, there has long been a concern to demonstrate the subject's relevance (Foster 1969; Grillo 1985; Sillitoe, in press). Some colonial administrations appointed government anthropologists, such as Francis Williams in Papua and Robert Rattray on the Gold Coast, who accomplished some significant work, although it was largely ignored by the academic community. They sought to convey something about local conditions to the authorities, informing officials such that their decision-making was more enlightened. Attempts to apply anthropology continued along much the same lines for the twentieth century, advising those trying to change other societies, for what they judged to be the better, seeking a role in the top-down imposition of development by informing administrators of probable social implications. Although they undertook some useful work, anthropologists turned increasingly to theoretical issues. An ambivalence becomes evident, even disenchantment with the idea of applied anthropology, which was to endure for the best part of the twentieth century, reflecting the emergence of an unhelpful divide between 'pure' and 'applied' anthropology. This has yet to be overcome fully, but the coming together of ethnobiology and applied anthropology is, I argue, making a noteworthy contribution to its resolution. Opinion was to harden even further with the independence of many colonies, leading to a welter of post-colonial criticism, while a sense of angst at being labelled the 'handmaiden of colonialism' prompted the subject to distance itself further. On the whole, those working in development were not particularly interested in listening to anthropologists anyway, who were often viewed as a nuisance, coming forward with awkward observations, even perceived as wishing to thwart projects. It was not easy to apply anthropology in such a climate.

Meanwhile ethnobiologists were toiling away in their corner of the sprawling discipline of anthropology. The first studies, closely related to, or conflated with, ethnoscience and ethnoecology, include some that take up in a more systematic fashion the work of the earlier naturalists who showed an interest in local plant and animal lore. They commonly involved the compiling of lists of native plant and animal resources, with local and scientific names, and include accounts of what people used these for, characterized by Conklin as what might 'be said to treat of botany [or zoology] with notes on ethnology' (quoted by Berlin 1992: 4). Not only ethnologists but also biologists contributed to this work, such as the botanist Harley Bartlett, whose work with the Batak in Sumatra marks him out as the 'grandfather of modern ethnobotany', according to Berlin (1992: 54). By the mid-twentieth century these utilitarian-focussed studies started to give way to more cognitively framed ones, notably studies that centred on elucidating classificatory schemes. A large part of this work has sought to establish universal principles that underlie human classification of nature. It claims to identify surprising parallels around the world between folk biology and the hierarchical taxonomy of biological science. Others have questioned the universality of this way of structuring understanding of the natural world.

This cognitively focused work became entangled with what at the time was boldly called the 'New Ethnography', which claimed to evolve a methodology that allowed researchers to access the native mindset without distortion and record the local view of the world, largely through close attention to semantics and positing a clear distinction between the local view – the emic perspective – and the theorists' view – the etic. As Frake put it, 'We must get inside our subjects' heads' (1964: 133). The subsequent postmodern critique has revealed their inadequacies. None the less, some impressively thorough investigations were undertaken, albeit often of narrow topics. According to Berlin

(1992: 36), in his well-informed overview of ethnobiological classification, the papers by Frake (1961) and Conklin (1962) on Subanun disease diagnosis and Hanunóo classification of peppers, respectively, were particularly influential in ushering in the new cognitive focus of ethnobiology, which was to dominate for the remainder of the century.

These studies are about as far removed from applied anthropology as is imaginable; as Ellen observes, 'the writing was scholarly, esoteric and fundamental' (2002: 237). He cites the example of Metzger and Williams's (1966) account of Tzeltal firewood classification, panned by Berreman (1966) as 'science of trivia' and parodied, along with Frake's and Conklin's work, as 'anaemic and emetic analyses'. But as Ellen points out, few today would think that an interest in how people select firewood is a trivial topic with the fuel crisis that many face, such as those living on the floodplains of Bengal and in Africa's savannah regions (Leach & Mearns 1988). This is no isolated whimsical example but stands for what was to happen with respect to ethnobiological studies in the future as trivia became treasured intelligence, and confirms the wisdom of Evans-Pritchard's (1946: 94) observation that to allow practical concerns alone to dictate research agendas is foolhardy.

The development connection: ethnobiology and indigenous knowledge

In addition to anthropologists, researchers from other disciplines such as botany, zoology, agriculture, and ecology continued to have an interest in ethnobiology, carrying on the earlier tradition. They represent a ongoing concern for practical issues, maintaining a focus on the utility of knowledge, and showing little interest in debates on human cognition. Recently these academic and practical strands have shown signs of coming together under an applied anthropology agenda, informing one another in an increasingly productive interdisciplinary interaction. The impetus for this has come from outside anthropology with the change in international development from a top-down to bottom-up approach, an emphasis on sustainability, and focus on poverty alleviation. Ethnobiology suddenly finds itself relevant to tackling development problems, as much of its work has direct relevance to those working in development, notably in the fields of natural resources management, environment, and health, and is seen particularly to have something to offer to currently popular calls for sustainable development – unlike some anthropology, with its introspective concerns for identity, selfhood, and so on.

The opportunities for anthropology to contribute to this work have perhaps never been better, with the acceptance of people's participation by mainstream development in the last two decades or so, and, more recently, associated with this participatory trend, the realization that so-called 'indigenous knowledge' has something important to contribute to development (Antweiler 1998; DeWalt 1994; Kloppenburg 1991; Purcell 1998; Sillitoe 1998a). Some anthropologists with ethnobiological interests need to look beyond their narrow concern with cognitive taxonomic issues, and focus more on how people use their knowledge in their lives to capitalize on the applied opportunities. The uses to which people put their knowledge of the natural world, and classification as it relates to their exploitation of it, are the aspects of ethnobiology that attract development personnel – being interested in the potential and exploitation of natural resources, they are particularly attentive to how people use their environment, in addition to how they talk about it.

Ethnobiology has reason to be proud as the sub-disciplinary home of many of us who have engaged with indigenous knowledge in development. Indeed it is arguable that ethnobiology, together with human ecology and farming systems research, contributed to this shift in development practice by demonstrating the depth of people's knowledge of their environments, contributing to the scotching of the pernicious 'ignorant peasants' myth (Sillitoe 1998b). Many ethnobiologists now contribute to sustainable participatory development initiatives – a considerable proportion of the papers presented at the 9th International Congress of Ethnobiology in 2004, for instance, showed an interest in indigenous knowledge and development issues. This engagement raises some important issues as to exactly what we are doing. One manifestation is the furious, if 'arguably futile' because 'ultimately irresolvable' (Ellen 2002: 236), debate over the propriety of the many proposed terms jostling for 'political correctness' which has distracted us from the task of indigenous knowledge research (Sillitoe 2002a).

There are various ways one can envisage applying anthropology and ethnobiological knowledge, but all pose challenges that we need to debate. I propose to explore five possible features of the application of anthropology together with ethnobiology, leaving on one side the applied anthropology that seeks to advise about the social implications of development interventions, with its focus on institutions, which has attempted with varying degrees of success over the years to engage with administrators and development policy-makers, and has evolved largely into current-day social development. The ethnobiologically informed approach to applied anthropology opens up new ways to become involved in, and influence, development agendas, beyond the long-tried and often frustrating social route.

Assisting the introduction of exogenous technology
One way to apply anthropology is to better facilitate others' use of scientific technology. It is this potential that is of particular interest to development agencies, as this in essence is what much development is supposedly about.

Where anthropology might offer assistance, as seen in much indigenous knowledge work, is attempting to inform development personnel about lay knowledge, and vice versa, such that technical development interventions better fit the local situation and promote meaningful participation. It is similar in some respects to previous attempts to apply anthropology, seeking to inform outsiders, but it differs with its focus on negotiation with local knowledge as opposed to advising on possible social institutional problems, and in taking advantage of current opportunities for this bottom-up perspective to influence programmes, as opposed to seeking openings to ameliorate top-down impositions. We all know of too many examples where development programmes have degraded, not improved, the lives of the poor through plain ignorance of their lives, needs, and so on. We seek to navigate cultural differences that varyingly condition expectations. The challenge is to see that development practitioners appreciate the importance of cultural context, that they are sympathetic to how it can condition knowledge and practices.

Work undertaken on the Bangladesh floodplain with the UK Department for International Development and UNESCO illustrates how we might seek to facilitate the more effective targeting of development resources, focusing on the compatibility of local with scientific ideas, exploring ways to improve natural resources research by combining scientific study of natural resources with farmers' and fishers' local knowledge of resources.[1] We have sought to facilitate communication between scientists and local people on the assumption, fundamental to development interventions, that science may have something to offer them in tackling their problems. Furthermore, it is possible that if scientific and indigenous knowledge are comparable, and if scientists are able to access local knowledge, this might save on expensive scientific research – on the grounds that sharing what the local people already know may reduce the need to conduct research into some topics – and also facilitate empowerment of the poor – on the grounds that if their knowledge features prominently in any development initiative, this will give them a meaningful role in its planning and implementation. Others have attempted similar work (see Antweiler 1998; DeWalt 1994; Richards 1985).

We have compared local Bengali farmers' soil classification with that of soil scientists to explore parallels and differences, correlating the mapping of local soil names with a scientific soil survey (Sillitoe, Barr & Alam 2004). We assumed that farmers' knowledge of the soils in their fields is the most locally relevant understanding of those soils, and that furthering scientists' appreciation of this knowledge should assist them in their work on the local farming system. It was an eye-opener for scientists, for example, to learn of the variability in what farmers had to say about what they thought was the same soil, suggesting that single targeted interventions could be inappropriate, particularly when we considered differences in socio-economic position too. We were also interested to see if there are potential efficiency gains over expensive land and soil surveys in collecting and using local soil knowledge. We sought not only to evaluate understandings of soil distribution but also to assess the extent to which a local population's knowledge of its soils might substitute for, or complement, an expensive scientific soil survey. This would reverse the usual dialogue in development, by emphasizing local people informing scientists, assessing the extent to which they might communicate intelligence about their soils, so reducing the need to undertake costly soil survey work, in addition to facilitating the communication of locally perceived problems.

In other work we have compared and contrasted aspects of scientists' knowledge of fish with that of people living on the Bangladeshi floodplains, for whom fishing is a significant source of food, reviewing local and scientific classifications of fish, and contrasting the use that local people and scientists make of their respective schemes (Alam & Sillitoe 2003; 2004). We noted what local people think are important features in characterizing different fish, notably why they value some more highly than others (e.g. good to eat, tasty, cheap to buy, easy to catch, undemanding to keep in ponds, cook well, etc.), and compared these ideas with those of fish scientists, notably the fish species they research and why they work on these fish and not others (i.e. they grow large, grow quickly, live in certain places, are easy to raise, breed easily, fingerlings readily obtainable, donor projects push them, etc.). We analysed these data according to local people's socio-economic positions (e.g. rich farmers, fishers, sharecroppers, labourers, etc.), gender, the geographical locale of their villages, their religion (Muslim, Hindu, etc.), and age. For fish scientists we analysed the data according to people's position (e.g. university lecturer, researcher, aid worker, donor official, etc.), gender, whether they were Bangladeshi or foreigner, their home town in Bangladesh (or elsewhere), and age.

We were interested not only in investigating differences between local people's ideas about fish and scientists' ideas, but also how local people's ideas differ from one another and how scientists' ideas differ from one another (e.g. by sex, occupation, etc.). We have drawn comparisons with studies conducted on rice-farming, where plant breeders have developed new high-yielding varieties that are unpopular with many people because of their poor organoleptic qualities and the expense of cultivation, with poor farmers having difficulty cultivating them because of the cost of inputs, and so on. The domain of crop-breeding is another where anthropologists have made noteworthy contributions to the incorporation of local views in the development agenda, such as Paul Richards's (1985) work in West Africa and David Cleveland and Daniela Soleri's (2002) in Meso-America. Our objective in the case of fish knowledge in Bangladesh was to assess the extent to which scientists' work in fisheries and aquaculture development interventions (to breed larger fish, to find better ways to feed pond fish, to promote rapid fish growth, to discover how to stop fish diseases, and so on) match local people's needs and ideas. A related issue is that the promotion of certain fish species, some of them exotic, over others has had an impact on the local fish populations and ecology. Some local species find themselves at a competitive disadvantage. The environmental changes resulting from the extensive flood prevention engineering schemes in Bangladesh have further changed aquatic resources. A focus on local knowledge of fish populations and ecology should reveal important intelligence pertinent to conservation to at least better understand, if not reverse, these changes.

This sort of work presents us with several challenges. Linkage of local knowledge and practices to the scientific technology that underpins development programmes is not straightforward. It can make the technicians' jobs more difficult by raising unanticipated complicating issues, albeit ones that they should wrestle with if they intend to further sustainable change, not proposing solutions to problems that they define, which regularly falter when projects finish. Anthropologists have to be careful not to encourage their reputation as the 'awkward brigade' of development consultancy, those who repeatedly create problems for others, leading to antipathy towards us and a reluctance to listen to what we have to say, which has largely happened throughout the discipline's applied history (Wright 1995). It is necessary to think positively in presenting unexpected information to technocrats and administrators, whose jobs are to promote change and not engage in endless debates.

We have to anticipate problems communicating with scientists in such ethnobiological 'indigenous knowledge' work (Dixon, Barr & Sillitoe 2000; Sillitoe 2002b). While we found that scientists were willing to hear about ideas that fitted with their theoretical understanding of their domains, they were not interested in information that did not, which seemed irrelevant to them. The workhorse of science can be shown the trough of local knowledge but it cannot be made to think on it. Soil scientists were interested, for example, to match up local Bengali names for soils with their survey series and note the diagnostic features of relevance to farmers, management issues, and so on. But they could see no relevance to farmers' ideas about the soil menstruating at intervals, which sounded wrongheaded, whereas anthropologists find such information intriguing and meriting further inquiry. While the idea seems strange to most soil scientists, as it is so far from their knowledge of soil processes, it is one that affects farmers' soil management and they need to consider it, as it may influence their response to interventions to enhance soil fertility. It serves to illustrate the implications of a holistic approach, leading us on to ritual affairs through an association with the river Ganges deity Gonga, relating these ideas to the annual swelling of the river and bursting of its banks to flood large regions, with significant implications for silt deposition and soil fertility, but for soil scientists this is an abstruse way of considering the issue of fertility management.

This is relatively uncontroversial information of a technically related sort that development practitioners may find helpful when they can see some relevance to their work, and in which they often show considerable interest, in our experience. The problem is that it can soon merge into information that could be used to the detriment of local populations, such as details of the supply of chemical fertilizers to farmers that may feature some black market involvement necessary to the viability of their precarious enterprises. In other words, we become caught up in political issues.

Facilitating local solutions to development

Another way to apply anthropology is to support the use of other's knowledge to further development. This is to reverse the previous view of applied anthropology, considering not what outsider knowledge may have to offer adjusted to socio-cultural context but what insider knowledge may have to suggest about advancing development and coping with outside influences. It also relates to the issue of supporting people's use of local plant resources, such as for medicinals, where these are cheap and as appropriate as imported alternatives. We face something of a dilemma here because if local knowledge has development relevance, then we might assume that people should see this for themselves and it is for them to decide what might be drawn upon in any intervention, or to develop themselves, otherwise it is interference, of a kind that makes many anthropologists understandably eschew the idea of applying anthropology in any context. I believe that they are wrong, given the dreadful problems of poverty that many people endure. What we should be doing is interfering with the interferers. We have to think how we can apply anthropology to assist people in representing themselves without presuming to speak for them. Participatory issues are central here, ensuring that local people significantly take part in decision-making processes, such that they can decide what information should be divulged and what retained – and, in the latter event, that they have an opportunity to influence events without necessarily revealing all of their reasons.

These issues have come to the fore in conservation and biodiversity management, where it is now widely accepted that local communities should play a central role; a number of ethnobiologists have contributed to such programmes (Laird 2002; Posey 1999). Agencies are coming to realize that cultural diversity, local people's knowledge and practices, contribute significantly to our understanding and protection of natural environments. The World Heritage Convention of UNESCO recognized a new category in 1995, 'The Cultural Landscape', which acknowledges the complex relationships between humans and nature in the formation and protection of landscapes. In this context UNESCO and the UN Environment Programme approached me in 2004 about a consultancy investigating the linkages between cultural diversity, the environment, and human well-being, to report to the Global Ministerial Environment Forum with a view to the promotion of 'the conservation and sustainable use of biological diversity' in the context of the 'promotion of the fruitful diversity of cultures' (Terms of Reference). Here we see that the indigenous knowledge agenda relates not only to the issue of knowledge transfer but also to the need to preserve the diverse knowledge and practices that inform the management of local resources.

We are on the way to winning the battle to convince agencies that the idea of the pristine reserve protected from human beings is a romantic fiction because there can be few regions of the world other than the polar extremities[2] where humans have not intervened in natural environments and their activities have not become an integral part of the management of the environment as it currently exists. Nevertheless, there continues to be resistance to the idea of including local people in managing parks (as evidenced at the 5th World Parks Congress in Durban – Brosius 2004 – and the current dispute in Kenya over the rights of Okiek [Dorobo] hunter-gatherers – A. Kassam pers. comm.). A well-known example is the game parks of Africa, where the exclusion of the local pastoral people and small hunter-gatherer groups (Kassam & Bashuna 2004) resulted in dramatic changes in the local ecology, undermining the park environments that the authorities intended to protect, because the activities of the pastoralists had served a significant management function regarding waterholes and rangeland. It is hardly surprising that human activities are central to the environment as anthropoid creatures have been living in the region for some five million years. The pastoralists living there currently have several conservation practices, for instance with respect to woodland around streams and springs (Smith, Meredith & Johns 1996). While the international scientific community may identify some conservation priorities through survey work, and so on, we need to involve the local population in establishing many locally important species and ecosystem issues (Shiva 1994). These are the people who know the local environment, they are the experts regarding biodiversity matters, and it is necessary to involve them as it becomes evident that biodiversity conservation cannot rely on protected areas only (such as the Ngorongoro Conservation Area and Serengeti National Park, which are proving difficult to manage and police), but must also include areas where people live and work.

Furthermore, this will help to ensure that any conservation strategies take account of local needs and socio-cultural context. Organizations have emerged to assist local people to contribute to the determination of conservation strategies for their region, such as the non-governmental organization Korongoro Integrated Peoples Orientated to Conservation, which is seeking to advance understanding of the conservation measures inherent in Maasai land use and resource management, and give them a regional voice. As Smith and Meredith note, 'Biodiversity conservation is an activity that must recognize the legitimacy of human actors in the landscape and must acknowledge the absolute importance of local involvement in and support of conservation strategies' (1999: 375). It is arguable that African culture is imbued with a conservation ethic. As Getui notes, 'in the African context, harmony and well-being of one and all are stressed through kinship systems which extend to cover animals, plants and non-living objects' (1999: 456). We find expression of these ideas in totemic beliefs and associated supernaturally sanctioned taboos on the consumption of certain animals and plants. In this regard, conservation achieves a religious significance for African people; it is deeply ingrained into their lifeways.

However, unlike the lions, elephants, and antelopes that inhabit the parks, which evidence the same lifestyles generation after generation, the human beings change rapidly. Today they wish to travel by vehicle, hunt with telescopic-sighted rifles, own radios and mobile phones, and these technological interventions, manifestations of wider changes in their lives – opportunities to earn a cash income, improved health and growing populations – are changing their relationship with the environment (Fratkin 1997). Consequently, they are seen again as a threat to the ideal of the game park that attracts tourists. But we cannot preserve some version of the 'traditional' pastoralist in aspic practising unchanging management strategies of the environment and having an unchanging relationship with the local ecology. We know that all communities are continually changing, formulating new ideas, and revising understanding of issues, and we have to devise further sophisticated approaches to promote understanding of such fluid knowledge.

There is dissent in local communities as to the extent to which they should manage their lives to promote tourism through the game parks, a source of desirable income (to buy rifles, radios, etc.); some persons have other ideas, the most extreme being poachers (a lucrative occupation if you can get away with it). Often there is no clear community view, with people everywhere disagreeing to some extent and not always able to agree a consensus. How can we manage the resulting conflict? This relates to the problem of what right anthropologists have to present the ideas and practices of another community, let alone represent them in any discussions. There is much work to do on participatory methodologies (e.g. ensuring fair representation of various local views, working out procedures to contain conflict when views diverge, overcoming the manipulation of these methods by outsiders seeking to control people, etc.). Strangely enough, anthropologists have so far had a relatively minor hand in the development of participatory methodologies, despite the fact that these herald profound changes in the way they undertake research and offer unparalleled opportunities to take part in development; their slow engagement with participation has something of a late-closing stable doors feel to it.

Further problems arise with the privatization of land resources, seeking to confine families to small ranches, undermining clan-based communities and the way they manage their environment (Fratkin 1998). Savannah pastoralists need to be able to move their herds over considerable distances to secure pasture and water throughout the year, and confining them to limited areas and out of reserves (as well as excluding them from areas designated as cultivable) undermines their subsistence and results in significant environmental change. The promoters of such policies have erroneously drawn on the 'tragedy of the commons' caricature of resource rights, an argument that relates not to common property rights of the kind that regulate pastoralists' use of land but to open access where no social regulation of land use is apparent (Hardin 1968; McCay & Acheson 1987; Niamir 1995; Spencer 2004). One problem is that we cannot turn the clock back and dissolve, for example, the states of Kenya and Tanzania, which would be necessary to allow the Maasai to reassert control over their territory along previous tribal lines. They have to go forwards from where they find themselves today, members of nation-states that do not always have their interests uppermost in mind when engaging in 'development', and desirous themselves of the benefits they perceive that commercial engagement with the wider world will bring them. But it is probable that involving them more closely in endeavours to ensure conservation of their region's biodiversity will pay dividends, drawing on their heritage in ways acceptable to them, be it eco-tourism, sustainable ranching, or whatever.

Furthering cultural diffusion
Ethnobiological applied anthropology could extend to establishing whether people's knowledge and practices in one place have relevance somewhere else. This is to exploit anthropology's comparative tradition in a novel way, and offers an innovative approach to development, as we have hardly explored the ethnographic record for such possible cross-cultural applications (in this regard Hill 1985, for example, suggests improving our classification of agrarian systems). It suggests a genuine application of anthropology and presents some real opportunities. It offers the prospect of using the extensive ethnographic corpus that is anthropology's legacy, which it has the necessary skills to retrieve

and interpret. In a sense it is to facilitate diffusion, which is what development is about essentially, except that development seeks to promote it in a limited one-way direction, from the West to the Rest. We are considering diffusion among the Rest, in seeking to establish whether knowledge and practices in one place may have relevance elsewhere. But without the full participation of the original knowledge-holders this could amount to the misappropriation of their cultural heritage (see later). Again we see the need to evolve research methods that meaningfully include all interested parties. It will be helpful to specialize in particular fields, as ethnobiologists customarily do, to understand the relevant issues in interdisciplinary contexts and have a ready familiarity with the literature. This will involve skill in navigating the pitfalls of cultural difference, aware that practices set in different cultural contexts may condition their expression in ways that are often difficult to predict. We must beware of rushing fences here. This is a message that the indigenous knowledge initiative sends constantly to the development community. We are not just talking about knowledge of soil management or whatever; wider cultural context is critical.

Some NGOs, such as Practical Action (previously the Intermediate Technology Movement – Practical Action n.d.), have undertaken such work and could gain from anthropological support. Libraries and archives hold a large amount of knowledge, some of it now forgotten by its original owners. There is opportunity here for methodological advances, such as the advancement of interactive ethnographic databases and e-science developments that will comprise the next generation of the Human Area Relations Files. Some institutions are already setting up such digital databases, such as the Council of Scientific and Industrial Research in India, which is creating a database to house India's 'traditional knowledge' that will include both ancient texts and current practices (Jayaraman 1999; Mashelkar 2003). These digital developments present opportunities and dangers, as databases may empower people or sideline them, depending on how they are deployed (Barr & Sillitoe 2000), such that we need to consider not only the technical issues involved in setting them up (Walker, Sinclair & Kendon 1995) but also the political aspects of their management (Agrawal 2002). It is political issues that are driving the establishment of some databases, such as the Indian one, to protect indigenous knowledge from commercial theft by confirming 'prior art' (see below). Now that the technology has arrived, we have to come up with ways to use it to good effect. We should involve ourselves, as we are well placed, with our cross-cultural awareness, to play a useful role in furthering the use of databases to facilitate the sort of ethnographically managed diffusion to inform development described here.

The issues of food security and sustainable farming are prominent concerns in development, and governments and agencies have invested large sums in agricultural research. The development assumption is that science may come up with advances to reduce hunger and land degradation. But it is equally probable that people in one region of the world may follow farming practices that could form the basis of innovations and adaptations elsewhere: for example, that tropical forest cultivators in the Pacific, South America, and Africa might have something to learn from one another's farming practices and innovations. In other words there could be a role for persons other than, as one Sunday newspaper expressed it, 'smart outsiders sent in to solve the problems of the world's poor', breeding new higher yield varieties or possibly today genetically modified crops, together with recommending synthetic inputs, notably fertilizers and biocides. Malnutrition is prevalent in large areas of Africa, with famines a too regular feature of news bulletins, which is attributed in part to the poor soils that characterize this ancient land mass that have proved particularly intractable to efforts to raise their fertility and productivity. Similar soils – Oxisols – are found in lowland South America, and until recently it was assumed that the mobile populations there were obliged to subsist through a mixture of slash-and-burn farming and forest foraging, a system made possible by their low population densities. The destruction of vast swathes of forest by *caboclo* settlers and commercial ranchers, degrading the resources of large areas, was taken as further evidence of the intractable nature of these heavily leached yellow-red soils that could not apparently support large-scale production and demand careful management.

Recently, however, considerable interest has been shown in areas of black soil *terra preta do Índio* which are found throughout non-flooded parts of Amazonia, some of these extending over large areas up to 500 hectares. These soils are the result of human activity, as evidenced by remains such as potsherds, their occurrence in upland places where human activity is the only conceivable depositional process, and high charcoal content resulting from burning of vegetation. Archaeological analysis indicates that these soils date from 360 BC to 1440 AD, that is, up to the Portuguese invasion, when the historical evidence suggests the population crashed, largely due to introduced diseases against which people had little immunity. This evidence has contributed recently to a revision of human-environment relations in the region, which were previously thought to be tightly constrained by the highly weathered and acidic soils limiting crop productivity and obliging people to practise shifting cultivation and forage in the forest. Adopting a historical ecology approach suggests that contemporary Indian subsistence practices were a response to the devastating impact of European contact and resulting depopulation (Balée 1998). Previously, it is suggested, populations were more sedentary and even able to cultivate areas intensively. The implications are considerable, not only for farming in Latin America but also for other tropical regions with similar soils such as Africa, and they are spurring further research.

How did people form and maintain these rich black soils? The evidence suggests that they are the result of a combination of midden deposit, regular burning, and mulching regimes (McCann, Woods & Meyer 2001). The ash and charcoal changed the soil chemistry, decreasing aluminium levels, increasing cation retention, and encouraging the development of a rich soil microbiota. Current research suggests that the secret to their formation is to burn vegetation partially, not reduce it all to ash, and combine this charcoal-rich mix with the soil (Prance 2004: 33). These microbiologically rich soils show a remarkable regenerative ability: farmers can even move them to new locations and they will reproduce. Contemporary *caboclo* farmers find them easy to manage sustainably, with appropriate rotations and inputs, although like any soil they are subject to degradation in careless hands. It is not suggested that we might recommend the transfer of *terra preta* establishment and use to Africa without giving careful consideration to the different environmental and socio-cultural conditions, such as the implications for labour demands, suitability to preferred crops, and so on. But this anthropogenic soil merits further investigation not only to intensify Amazonian farming and reduce the ephemeral cultivation so destructive of the rainforest, but also as a technology that could benefit other regions with such weathered soils. These black soils are particularly popular among Amazonian farmers for higher value marketable crops that mature quickly and are harvested before weeds become a serious problem (German 2003). They are less favoured for some longer-growing staple crops, notably manioc, which farmers report yield better on the poorer red soils, underlining the need to consider cropping regimes – *terra preta* encourages prolific growth of leafy vegetation at the expense of the harvested roots, in addition to supporting more vigorous weed growth that demands laborious clearance.

Some places in Africa do not have ancient highly weathered soils but new ones derived from volcanic materials. These soils – andosols – contrary to popular opinion, are not particularly rich in available nutrients and present farmers with fertility problems, notably with phosphorus, which they fix ferociously. In the mountains of New Guinea there are people who farm similar soils in an intriguingly sustainable way. Although people here may move garden sites and clear new cultivations when crop yields fall below certain tolerable levels, in some locales they keep gardens under cultivation for decades, even generations (Sillitoe 1996). So long as yields are acceptable, and garden location convenient, they may continue cultivation indefinitely. The soil properties that limit crop production, besides low available phosphorus, are acidity interfering with nutrient supply, depressed cation availability (particularly potassium), and high organic matter reducing nitrogen availability (Radcliffe & Gillman 1985). Soil physical properties are good by contrast. In the initial clearance all vegetation of suitable size is burned, except for that used for fence stakes or log barricades. Nutrients locked up in the vegetation are rapidly returned to the soil in a readily available form via the ash. Controlled burning gives a critical, though short-lived, boost to the availability of several elements contained in the vegetation, notably phosphorus, potassium, and nitrogen, and increases pH, further promoting the availability of critically limiting nutrients sufficient to permit cultivation of a variety of crops. This crop diversity is short-lived, paralleling the ephemeral nature of improved nutrient availabilities. Sites cultivated repeatedly support sweet potato largely. This crop occupies a central place in the near-continuous farming system throughout the New Guinea highlands. The changes that occur in soil fertility status with time under cultivation do not necessarily reduce yields; contrary to expectations, farmers maintain that the soil may improve with use, gardens experiencing an increase in staple production.

How do they do it? Local soil management featuring soil mounds enables nutrients stored in plant residues to be incorporated into the soil during cultivation (Waddell 1972). When a plot is re-cultivated and planted, weedy re-growth and crop residues are composted into earth mounds, or, if left fallow for a longer period, coarse grasses and herbs are uprooted and incorporated, either as compost or as ash (Floyd, Lefroy & D'Souza 1988). A remarkable feature of this semi-permanent farming system is that, other than planting material, nothing comes from outside the sites during their entire productive life-cycle. The regime does not lead to a decline in productivity as expected due to nutrient losses, weed proliferation, disease build-up, or soil depletion through erosion. The mechanism of nutrient uptake that mounds afford is effective at overcoming phosphate fixation and poor base saturation. The boost in available potassium from grasses in the compost is also important to the success of mounds in sustaining long-term cultivation. Mounding ensures that the soil is friable, the compost providing a soft centre into which tubers can readily swell. It also increases topsoil depth. The incorporation of plant material increases microbial populations and decomposition rates, and burying weed residues reduces germination and gives the crop a head start. The decomposing compost also increases internal temperature and improves water-holding capacity. It reduces disease and rotting of tubers too. There is little chance that local farmers can significantly increase their composting rates because of the limited availability of suitable plant material, and the labour demands to collect it. If people collected material from elsewhere, this would deprive other fallowed gardens of compost, undermining their long-term fertility. None the less, this mound and compost method of soil management deserves attention; it could have some applicability to other regions with Andosols, such as parts of East Africa. In northern Rwanda, for example, where similar farming practices occur, people manage 'weeds' as nutrient stores to be released at appropriate moments in the cultivation cycle, often with the burning of plant residues, and build composted mounds to cultivate crops that include sweet potato, beans, maize, and bananas (Fairhead 1990). We have much to learn about the variety and dynamics of such tropical agricultural systems and what these farmers could teach one another.

Advancing the commercial use of knowledge

Another possible interest in applied ethnobiology is in reversing the usual direction of knowledge flow in development contexts, in other words to conduct research elsewhere to find intelligence that may be useful to Western societies. Sometimes researching other knowledge traditions turns up intelligence new to science, which it subsequently incorporates, continuing a long tradition mentioned earlier. Some engage in ethnographic inquiries explicitly to find commercially exploitable knowledge, as in bio-prospecting, which blatantly seeks to profit from others' heritage (Greene 2004). Such researchers work with local populations in the hope that they know something that will prove of benefit to Western societies, and that they will return with trophies such as a plant extract that can be used in treating cancer effectively (Svarstad & Dhillion 2000). The majority of anthropologists are uneasy with such aims, if not implacably opposed to them, since those funding such work are out to make a profit from others' knowledge.

We enter the contentious legal tangle of intellectual property rights, as populations try to protect themselves and seek fair play in any application of their knowledge (Dutfield 1999; 2003; Posey 2000). The exploitation of their knowledge has made some tribal people suspicious, even unsympathetic, with regard to any research in their communities, such as many groups in Amazonia that seek to protect their knowledge by vetting researchers closely or prohibiting them entirely. Anthropologists might act as advisers here, or advocates according to some, both to communities that find their knowledge and resources subject to commercial interest, and to outside companies that are looking to exploit them (Ervin 2000: 121-40; Grillo 1985: 25; Hastrup & Elass 1990; Van Willigen 1986: 111-25). Such involvement supposes both an acquaintance with the region ethnographically and some education in international commercial law and associated political policy environment – although potentially useful, persons with such knowledge are currently uncommon. While there are international discussions about creating legal procedures to protect the interests of local populations and thwart biopiracy, such as TRIPs (the Agreement on Trade-Related Aspects of Intellectual Property Rights), each population has its own cultural, historical, and environmental peculiarities, as any ethnographer is well aware, that no legislation can deal with entirely.

An intriguing current case involves a succulent plant (*Hoodia* sp.) found in South Africa that the San people have used for generations to assuage feelings of hunger and thirst – useful to hunter-gatherers in such an arid region. It is thought that a processed extract from the plant might retail as an appetite-suppressing drug potentially worth £2 billion as a medicine to control weight, particularly in Euro-America, where obesity is now rated a major health issue (Laird & Wynberg 2002; Wynberg 2004). The CSIR, a statutory South African scientific and technological research institute, has patented the plant's active constituents and signed a licensing agreement with Phytopharm, a company that specializes in phytomedicine promotion. Phytopharm in turn has licensed the drug company Pfizer to develop and market the drug globally. The drug is expected to generate millions in royalty payments if it successfully comes to market. Initially there was no arrangement for sharing these substantial benefits with the San, original knowledge-holders of the plant and its properties, which echoes the abuse of San land rights in the Kalahari by the Botswana government forcibly relocating them to camps, as highlighted in a current Survival International campaign (Survival International n.d.). But subsequently the San negotiated an agreement with the CSIR for a share of the royalties, which suggests that patents may serve the interests of indigenous knowledge-holders. While this may be taken as evidence that these people are as interested in engaging in market relations as Western capitalists, what such arrangements mean to them culturally is unclear, and how they might use any income for the benefit of their entire community has yet to be worked out, the plant knowledge generating the income being common property.

Benefit-sharing is a socially and environmentally complex issue and alternative approaches to commercial partnerships and ways to protect intellectual property may be more appropriate and merit exploration (Wynberg 2004). It is here that an anthropologist might find a useful role. There are precedents, such as anthropologists advising advocates of Aboriginal land rights in Australia (Layton 1985; 1997; Maddock 1998). Furthermore, such work needs to proceed not only case by case, sensitive to cultural and historical nuances, but also as part of a regional strategy to the commercialization of such ethnobotanical knowledge. In South Africa, for example, the CSIR has launched a major bio-prospecting project to seek for plants with commercially exploitable properties (see CSIR n.d.). It has a policy of working with a few local healers to 'take account of the rights, interests and practices of indigenous peoples' and to advance 'equitable collaboration agreements' (cited by Laird & Wynberg 2002: 64). The implications are considerable: for example, the companies Bioharmony Africa and Phyto Nova market products in health food shops in the United Kingdom that claim to be 'harnessing the healing power of African traditional medicine' (Bioharmony n.d.), their brochures going on to describe them as 'an innovative South African enterprise of integrity and credibility comprising a unique multi-disciplinary and committed team of medical doctors, scientists and indigenous doctors to harness the synergies between indigenous knowledge systems, scientific research, modern technologies and biodiversity conservation'. Worryingly, there is no mention of an anthropologist in the multi-disciplinary team, although another brochure assures the shopper that these companies aim for 'the support and upliftment of local disadvantaged communities'.

Similar examples can be multiplied from around the world. The extensive forests of the Amazonian region have proved attractive to those seeking new products for global medicine, cosmetic, and food markets. Pharmaceutical companies are on the look-out especially for alkaloids to extract and test for therapeutic potential. Shamans are particular targets as persons with detailed knowledge of plants' medicinal properties. The result is many partnerships between companies and local communities brokered by intermediaries such as the International Cooperative Biodiversity Group (n.d.), working, for instance, with villages in Surinam and the US drug company Bristol Myers-Squibb. The cupuacu (*Theobroma grandiflorum*), an orange-skinned, Amazonian forest fruit, a relative of cocoa, has recently attracted commercial interest both as a food, serving as a fruit juice, fruit tea, and jam ingredient (with particular nutritional benefits, being high in certain vitamins, fatty acids, and chocolate flavonoids minus theobromine), and as a moisturizer ingredient: cupuacu butter is being promoted as a cosmetic that promotes youthful elastic skin. Two Japanese companies – Ashai Foods and Cupuacu International – seeing the fruit's potential, have registered trademarks, to the anger of people in Brazil led by an NGO, Amazonlink, who say this is corporate biopiracy. This contrasts with the cosmetics firm Aveda, which is working in collaboration with the Yawanawa people of Brazil in developing its butter emollient (Zobel 2004).

In the mountainous Yunnan province of China we find the plant *qinghao*, an ancient traditional remedy for malaria mentioned in Chinese medical almanacs dating from 340 BC, but only recently 'discovered' by the outside world. The active ingredient, an extract called artemisinin, has been synthesized into a drug called Artekin, recently launched internationally with WHO support (Bodeker 2003). The property right issues here became entangled with the Cold War: the Chinese communist regime, fearing that Western pharmaceutical companies would steal artemisinin, kept it secret for many years after extracting it, having done so in part to assist Vietnamese troops in the war against America. In addition to advising on local cases where they have expertise, knowing that cultural context can significantly influence outcomes of property right negotiations, there may also be a place for anthropologists in the deliberations of international bodies on property rights, such as the WIPO's

(World Intellectual Property Organization) Report on Traditional Knowledge, the first of its kind to consider the intellectual property needs of 'traditional knowledge' holders, or the Traditional Ecological Knowledge Prior Art Database (TEKPAD) under the aegis of the Science and Human Rights Program of the American Association for the Advancement of Science, which links with patent offices in Europe and the United States and aims to protect indigenous knowledge against inappropriate patents on the principle of 'defensive disclosure', that is, establishing prior art status (TEKPAD n.d.).

Supporting alternative development
In addition to fights to protect others' rights (Jerome 1998), there are critiques of the idea of development as a capitalist concept that has been exported to the rest of the world, or imposed on it, depending on your view (Escobar 1995; Hobart 1993; Keen 1999: 40-7; Mosse 2005). Engagement with development implies not only contemplating technical matters, deploying ethnobiology with applied science, but also, as mentioned, confronting political issues, considering alternative views of human-environment relations. The political dimension is prominent at all levels where we challenge and seek to convince authorities that there are benefits to be gained by giving more opportunity to local communities to determine their own destinies in the light of their knowledge and values (Gough 1968; Huizer & Mannheim 1979; Stavenhagen 1971; Tax 1975). This leads to the view that applying anthropology should involve engaging in ethnographically informed criticism of development, not drawing on our ethnobiological knowledge to assist agencies in their current work, but instead lending support to communities to put forward their own alternative views of development, which may differ radically from those of capitalism. Anthropology should not allow development to incorporate parts of its practice as indigenous knowledge, adding them to its many other approaches. Involvement in the promotion of such counter-development implies engagement in global politics. The problem is making alternative views heard. Current global political arrangements make this doubtful, and, even if heard, they may be construed by the powerful as inimical to the world order. But in supporting communities to promote their views, there is again the danger of interfering in others' destinies, a possibility that prompts some anthropologists, as noted, to maintain that applying anthropology is wrongheaded. What role is there for an ethnobiologically informed applied anthropology when people wish, and are able, to represent themselves?

One way in which we might legitimately seek some role in the context of such counter-development is through supporting the visions of people elsewhere regarding their futures. The campaign to establish a meaningful place for indigenous knowledge in development is one context. While some of us work at the grass roots to further this, trying to get the local voice heard both by managers and by staff working on projects at the community level (as related above in the Bangladesh context), others have campaigned at the international level, seeking to lobby and influence politicians and policy-makers in agencies (Sillitoe & Bicker 2004). We can obtain some feel for the possible outlines of such alternative development agendas by considering the endeavours of indigenous activists at various international conferences seeking to thrash out conventions and agreements, resulting in a growing number of declarations by indigenous peoples' organizations (Indigenous Peoples' Organizations 1999). A recurring theme is that any hope of sustainable development is compromised while an affluent minority consume the larger part of the earth's resources and a poor majority struggle to survive. An increasing gap between rich and poor, together with burgeoning global environmental problems, such as ozone depletion and global warming, have led to the interlinking of sustainability and development, and several international conventions have involved indigenous people and stressed the value of 'traditional' ethnobiological knowledge, calling for the empowerment of local communities and protection of their land and resource rights.

In 1992, the United Nations Conference on Environment and Development – the so-called 'Earth Summit' – excluded minority groups and NGOs; these organized a parallel 'Earth Parliament' at Kari-Oca on the outskirts of Rio de Janeiro and lobbied delegates vigorously. The resulting Kari-Oca Declaration, or the Indigenous Peoples' Earth Charter, gives an alternative view of development and environmental issues.

> We, the Indigenous Peoples, maintain our inherent rights to self-determination. We have always had the right to decide our own forms of government, to use our own laws, to raise and educate our children, to our own cultural identity without interference ... We maintain our inalienable rights to our lands and territories, to all our resources ... We assert our ongoing responsibility to pass these on to the future generations (Indigenous Peoples' Organizations 1999: 559-60).

The Declaration demands self-development, recognition of rights, and respect for cultural heritage as necessary to sustainable development. The international community came to acknowledge that indigenous people must play a key part to make development sustainable. The Rio Summit furthered indigenous peoples' interests considerably, calling for nations to adopt appropriate policies and legal instruments, recognize indigenous values, knowledge, and resource rights, and promote local participation in sustainable development strategies. In the words of the Rio Summit Declaration (Principle 22): 'Indigenous people and their communities, and other local communities, have a vital role in environmental management and development because of their knowledge and traditional practices' (Tebtebba Foundation n.d.).

Action since Rio has been disappointing regardless of conventions and declarations of principle recognizing that indigenous peoples have a role in sustainable development, and translating words into action, from a local to an international level, remains an enormous challenge. Governments lack the political will to sign up, and agreements, few of which have been honoured adequately, reaffirm the sovereignty of nation-states over natural resources to be traded on international commodity markets. The dominance of international economic and financial bodies (such as the World Trade Organization, the World Bank, and the International Monetary Fund, which support so-called 'economic globalization') has undermined the Rio agreements on sustainable development and effectively sidelined indigenous peoples. In the words of the Indigenous Peoples' Caucus Statement (17 May 2002), issued in Jakarta in the lead-up to the Johannesburg World Summit, development should be

> about addressing social and power relationships, and about how these relationships impact on our relations with the Earth. The contemporary world is characterised by deep imbalances in our social relations, of gross inequalities between nations and within societies, manifested by huge disparities in consumption ... Governance structures for Sustainable Development must strive for greater democratisation, transparency, equity, and accountability in order to achieve better outcomes (Tebtebba Foundation n.d.).

The Johannesburg World Summit on Sustainable Development in 2002 was seen as another opportunity to turn rhetoric into reality. The recognition of indigenous peoples afforded them a role in Summit preparation to comment officially on implementation of commitments since Rio and on future priorities. Vigorous lobbying ensured that the Johannesburg Summit Political Declaration importantly stated: 'We reaffirm the vital role of the indigenous peoples in sustainable development' (Tebtebba Foundation n.d.). Indigenous peoples again held an alternative summit of the excluded concurrently, at Kimberley in South Africa, at which concern was expressed at the lack of progress during the previous decade to implement the Rio agreements. It agreed the Kimberley Declaration, which again articulates an alternative vision of development, reaffirming rights to self-determination and to control resources.

> We reaffirm our relationship to Mother Earth and our responsibility to coming generations ... Our lands and territories are at the core of our existence ... we have a distinct spiritual and material relationship ... they are inextricably linked to our survival and to the preservation and further development of our knowledge systems and cultures ... The national, regional and international acceptance and recognition of Indigenous Peoples is central to

the achievement of human and environmental sustainability. Our traditional knowledge systems must be respected, promoted and protected; our collective intellectual property rights must be guaranteed and ensured. Our traditional knowledge is not in the public domain; it is collective, cultural and intellectual property protected under our customary law ... We are determined to ensure the equal participation of all Indigenous Peoples throughout the world in all aspects of planning for a sustainable future with the inclusion of women, men, elders and youth. Equal access to resources is required to achieve this participation. We urge the United Nations to promote respect for the recognition, observance and enforcement of treaties, agreements and other constructive arrangements concluded between Indigenous Peoples and States (Tebtebba Foundation n.d.).

It amounts to a call to action, to support recognition of peoples' right to self-determination and representation through their own institutions, including the strengthening of their position with regard to knowledge-sharing. There is a need to increase support to communities to develop their knowledge and institutions, and to promote a development that respects others' identities. It is necessary to facilitate the control and management of lands and resources under customary arrangements, which many see as fundamental to poverty eradication, and which will require the resolution of contentious sovereignty conflicts. There is a need to insist on peoples' informed consent to any developments in their regions, which implies greater corporate accountability. What role should anthropologists seek in such campaigns? Again, to offer assistance demands some background in international law and policy procedures, in addition to anthropology. The boundary between giving advice and lobbying is indistinct and it is necessary carefully to consider the ethical implications of involvement, as in the end communities should reach their own decisions and represent their own interests. They may not want any well-meaning help from outsiders, previous experiences having made them suspicious of any such interference. While ideologically moved, but politically ineffective, intellectuals may encourage action research, this can soon become unwanted intrusion (Scheper-Hughes 1995).

Applying anthropology and ethnobiology

There are two conflicting views evident from this review: the present pragmatic and the future idealistic (Cochrane 1971: 9; Grillo 1985: 28-30), which reflect the 'pure' and 'practical' divide. The idealists criticize those who give in to the development establishment by co-operating with agencies. Idealists argue that this is wrong, that we should be seeking to topple them, not accede to their current fashions of participation and indigenous knowledge, which they think are in danger of being absorbed by development in such a way that it can continue with 'business as usual'. The pragmatists argue that the 'poorest of the poor' urgently require assistance and they are consequently doing their best to help agencies in their work to aid such communities. While they may also agree that we need to try to change the way in which agencies engage in 'development', they do not think that we should wait until this occurs before co-operating with them, as grinding poverty is a pressing contemporary problem. If ethnobiologists can reduce hunger and disease by becoming involved – under the guise of indigenous knowledge or whatever we wish to call it – then surely they should do so, as several are.

An unprecedented opportunity for applied anthropology has opened up with the arrival of indigenous knowledge in development, which has deep roots in ethnobiology. Whereas the previous applied anthropology with its focus on social institutions had to fit into a top-down framework, the current applied anthropology is able to focus on local knowledge systems in a bottom-up context more conducive to anthropological engagement. A rank academic outsider not so long ago, ethnobiology looks like a horse to back on this new applied course, following the advent of indigenous knowledge and focus on sustainable development. Its resulting association with applied anthropology takes ethnobiology way beyond its usual frame of reference. The future offers some intriguing challenges, as ethnobiology shifts from an academic to an action frame of reference within anthropology. However, we face a number of dilemmas when we leave the security of the academic seminar and engage with the problems of the real world. There are several ways to envisage the emerging combination of ethnobiology and applied anthropology, as outlined; some of them are questionable, all of them are problematic.

NOTES

For helpful comments on this paper I am grateful to Roy Ellen, James Fairhead and Aneesa Kassam.

[1] The ethnographic vignettes that I present in this paper all come from natural resources management as this is the field that I know best, but other domains such as health and medicine could equally well supply examples.

[2] It is arguable that even large parts of the Arctic and Antarctic, such as Spitzbergen and the Antarctic Peninsula, are markedly affected by human activities, as is the entire polar region with the melting of ice caps.

REFERENCES

AGRAWAL, A. 2002. Indigenous knowledge and the politics of classification. *International Social Science Journal* 54, 287-97.

ALAM, M. & P. SILLITOE 2003. Peasants versus ichthyologists: a comparison of indigenous with scientific fish knowledge in Bangladesh. Paper presented at the 5th Decennial Conference of the Association of Social Anthropologists, University of Manchester.

———— & ———— 2004. Local farmers' and fishers' assessments of, and adaptations to, the environmental changes resulting across the Bangladeshi floodplain with the construction of flood defences under the 'Flood Action Plan'. Paper presented at the 9th International Congress of Ethnobiology, University of Kent.
ANTWEILER, C. 1998. Local knowledge and local knowing: an anthropological anlaysis of contested 'cultural products' in the context of development. *Anthropos* **93**, 469-94.
BALÉE, W. (ed.) 1998. *Advances in historical ecology*. New York: Columbia University Press.
BARR, J.J.F. & P. SILLITOE 2000. Databases, indigenous knowledge and interdisciplinary research. In *Indigenous knowledge development in Bangladesh: present and future* (ed.) P. Sillitoe, 179-95. London: Intermediate Technology Publications.
BENNETT, J.W. 1976. *The ecological transition: cultural anthropology and human adaptation*. New York: Pergamon.
BERLIN, B. 1992. *Ethnobiological classification: principles of categorization of plants and animals in traditional societies*. Princeton: University Press.
BERREMAN, G.D. 1966. Anemic and emetic analyses in social anthropology. *American Anthropologist* **68**, 346-54.
Bioharmony n.d. Website (*http://www.bioharmony.co.za*).
BODEKER, G. 2003. The interface between traditional medical knowledge and modern science. Paper delivered at the British Association for the Advancement of Science Festival of Science meeting, University of Salford.
BROSIUS, P. 2004. Indigenous peoples and protected areas at the World Parks Congress. *Conservation Biology* **18**, 609-12.
CLEVELAND, D.A. & D. SOLERI (eds) 2002. *Farmers, scientists and plant breeding: integrating knowledge and practice*. Wallingford: CABI Publishing.
COCHRANE, G. 1971. *Development anthropology*. New York: Oxford University Press.
CONKLIN H.C. 1962. The lexicographical treatment of folk taxonomies. *International Journal of American Linguistics* **28**, 119-41.
CSIR n.d. Website (*http://www.csir.co.za*).
DEWALT, B.R. 1994. Using indigenous knowledge to improve agriculture and natural resource management. *Human Organization* **53**, 123-31.
DIXON, P.J., J.J.F. BARR & P. SILLITOE 2000. Actors and rural livelihoods: integrating interdisciplinary research and local knowledge. In *Indigenous knowledge development in Bangladesh: present and future* (ed.) P. Sillitoe, 161-77. London: Intermediate Technology Publications.
DUTFIELD, G. 1999. Introduction: rights, resources and responses. In *Cultural and spiritual values of biodiversity* (ed.) D.A. Posey, 505-15. London: Intermediate Technology Publications.
———— 2003. Legalized theft? Traditional knowledge and the patent system. Paper delivered at the British Association for the Advancement of Science Festival of Science meeting, University of Salford.
ELLEN, R.F. 2002. Déjà vu all over again, again. In *'Participating in development': approaches to indigenous knowledge* (eds) P. Sillitoe, A. Bicker & J. Pottier, 235-58. London: Routledge.
———— 2004. From ethno-science to science, or 'What the indigenous knowledge debate tells us about how scientists define their project'. *Journal of Cognition and Culture* **4**, 409-50.
ERVIN, A.M. 2000. *Applied anthropology: tools and perspectives for contemporary practice*. Boston: Allyn & Bacon.
ESCOBAR, A. 1995. *Encountering development: the making and unmaking of the Third World*. Princeton: University Press.
EVANS-PRITCHARD, E.E. 1946. Applied anthropology. *Africa* **16**, 92-8.
FAIRHEAD, J. 1990. Fields of struggle: towards a social history of farming knowledge and practice in Bwisha community, Kivu, Zaire. Ph.D. dissertation, London University.

FLOYD, C.N., R.D.B. LEFROY & E.J. D'SOUZA 1988. Soil fertility and sweet potato production on volcanic ash soils in the highlands of Papua New Guinea. *Field Crops Research* **19**, 1-25.
FOSTER, G. 1969. *Applied anthropology*. Boston: Little, Brown.
FRAKE, C.O. 1961. The diagnosis of disease among the Subanun. *American Anthropologist* **63**, 11-32.
——— 1964. Notes on queries in anthropology. *American Anthropologist* **66**, 132-45.
FRATKIN, E. 1997. Pastoralism: governance and development issues. *Annual Review of Anthropology* **26**, 235-61.
——— 1998. *Ariaal pastoralists of Kenya: surviving drought and development in Africa's arid lands*. Boston: Allyn & Bacon.
GERMAN, L.A. 2003. Historical contingencies in the coevolution of environment and livelihood: contributions to the debate on Amazonian Black Earth. *Geoderma* **111**, 307-31.
GETUI, M. 1999. Spiritual beliefs and cultural perceptions of some Kenyan communities. In *Cultural and spiritual values of biodiversity* (ed.) D.A. Posey, 455-7. London: Intermediate Technology Publications.
GOUGH, K. 1968. New proposals for anthropologists. *Current Anthropology* **9**, 403-7.
GREENE, S. 2004. Indigenous people incorporated? Culture as politics, culture as property in pharmaceutical bioprospecting. *Current Anthropology* **45**, 211-37.
GRILLO, R. 1985. Applied anthropology in the 1980s: retrospect and prospect. In *Social anthropology and development policy* (eds) R. Grillo & A. Rew, 1-36. London: Tavistock.
HARDIN, G. 1968. The tragedy of the commons. *Science* **162**, 1243-8.
HASTRUP, K. & P. ELASS 1990. Anthropological advocacy: a contradiction in terms. *Current Anthropology* **31**, 301-11.
HILL, P. 1985. The practical need for a socio-economic classification of tropical agrarian systems. In *Social anthropology and development policy* (eds) R. Grillo & A. Rew, 117-30. London: Tavistock.
HOBART, M. (ed.) 1993. *An anthropological critique of development: the growth of ignorance*. London: Routledge.
HUIZER, G. & B. Mannheim (eds.) 1979. *The politics of anthropology*. The Hague: Mouton.
INDIGENOUS PEOPLES' ORGANIZATIONS 1999. Appendix 1 Declarations of the indigenous peoples' organizations. In *Cultural and spiritual values of biodiversity* (ed.) D.A. Posey, 555-601. London: Intermediate Technology Publications (for the UN Environment Programme).
International Cooperative Biodiversity Group n.d. Website (*http://www.fic.nih.gov/programs/rfa.html*).
JAYARAMAN, K.S. 1999. And India protects its past online. *Nature* **401**, 413-14.
JEROME, J.S. 1998. How international legal agreements speak about biodiversity. *Anthropology Today* **14: 6**, 7-9.
KASSAM, A. & A.B. BASHUNA 2004. Marginalisation of the Waata Oromo hunter-gatherers of Kenya: insider and outsider perspectives. *Africa* **74**, 194-216.
KEEN, I. 1999. The scientific attitude in applied anthropology. In *Applied anthropology in Australasia* (eds) S. Toussaint & J. Taylor, 27-59. Nedlands: University of Western Australia Press.
KLOPPENBURG, J. 1991. Social theory and the de/construction of agricultural science: local knowledge for an alternative agriculture. *Rural Sociology* **56**, 519-48.
LAIRD, S. (ed.) 2002. *Biodiversity and traditional knowledge*. London: Earthscan.
——— & R. Wynberg 2002. Institutional policies for biodiversity research. In *Biodiversity and traditional knowledge* (ed.) S. Laird, 39-76. London: Earthscan.
LAYTON. R.H. 1985. Anthropology and the Australian Aboriginal Land Rights Act in Northern Australia. In *Social anthropology and development policy* (eds) R. Grillo & A. Rew, 148-67. London: Tavistock.

―――― 1997. Representing and translating people's place in the landscape of northern Australia. In *After writing culture: epistemology and praxis in contemporary anthropology* (eds) A. James, J. Hockey & A. Dawson, 122-43. London: Routledge.

LEACH, G. & R. MEARNS 1988. *Beyond the woodfuel crisis: people, land and trees in Africa*. London: Earthscan.

MCCANN, J.M., W.I. WOODS & D.W. MEYER 2001. Organic matter and anthrosols in Amazonia: interpreting the American legacy. In *Sustainable management of soil organic matter* (eds) R.M. Rees, B.C. Ball, C.D. Campbell & C.A. Watson, 180-9. Wallingford: CABI International.

MCCAY, B.M. & J.M. ACHESON (eds) 1987. *The question of the commons: the culture and ecology of communal resources*. Tucson: University of Arizona Press.

MADDOCK, K. 1998. The dubious pleasures of commitment. *Anthropology Today* **14**: 5, 1-2.

MARTIN, G.J., A.L. AGAMA, J.H. BEAMAN & J. NAIS 2002. *Projek etnobotani Kinabalu: the making of a Dusun Ethnoflora (Sabah, Malaysia)*. Paris: UNESCO People and Plants Working Paper 9.

MASHELKAR, R.A. 2003. How can we protect traditional knowledge in a global knowledge-based economy? Paper delivered at the British Association for the Advancement of Science Festival of Science meeting, University of Salford.

METZGER, D. & G.E. WILLIAMS 1966. Some procedures and results in the study of native categories: Tzeltal firewood. *American Anthropologist* **68**, 389-407.

MOSSE, D. 2005. *Cultivating development: an ethnography of aid policy and practice*. London: Pluto.

NIAMIR, M. 1995. Indigenous systems of natural resource management among pastoralists of arid and semi-arid Africa. In *The cultural dimensions of development: indigenous knowledge systems* (eds) D.M. Warren, L.J. Slikkerveer & D. Brokensha, 245-57. London: Intermediate Technology Publications.

POSEY, D.A. (ed.) 1999. *Cultural and spiritual values of biodiversity*. London: Intermediate Technology Publications (for the UN Environment Programme).

―――― 2000. Ethnobiology and ethnoecology in the context of national laws and international agreements affecting indigenous and local knowledge, traditional resources and intellectual property rights. In *Indigenous environmental knowledge and its transformations* (eds) R. Ellen, P. Parkes & A. Bicker, 35-54. Amsterdam: Harwood Academic.

Practical Action n.d. Website (*http://www.practicalaction.org*).

PRANCE, G.T. 2004. Indian black earth – *terra preta do Índio*. *Tropical Agriculture Association Newsletter* **24: 1**, 32-3.

PURCELL, T.W. 1998. Indigenous knowledge and applied anthropology: questions of definition and direction. *Human Organization* **57**, 258-72.

RADCLIFFE, D.J. & G.P. GILLMAN 1985. Surface charge characteristics of volcanic ash soils from the Southern Highlands of Papua New Guinea. In *Volcanic soils: weathering and landscape relationships of soils on tephra and basalt* (eds) E. Fernandez Caldas & D.H. Yaalon, 35-46. Cremlington: Catena Supplement.

RICHARDS, P. 1985. *Indigenous agricultural revolution: ecology and food-crop farming in West Africa*. London: Hutchinson.

SCHEPER-HUGHES, N. 1995. The primacy of the ethical: propositions for a militant anthropology. *Current Anthropology* **36**, 409-20.

SHIVA, V. 1994. (ed.) *Biodiversity conservation: whose resources? Whose knowledge?* New Delhi: INTACH.

SILLITOE, P. 1996. *A place against time: land and environment in the Papua New Guinea highlands*. Amsterdam: Harwood Academic.

——— 1998a. The development of indigenous knowledge: a new applied anthropology *Current Anthropology* **39**, 223-52.
——— 1998b. What know natives?: Local knowledge in development. *Social Anthropology* **6**, 203-20.
——— 2002a. Participant observation to participatory development: making anthropology work. In *'Participating in development': approaches to indigenous knowledge* (eds) P. Sillitoe, A. Bicker & J. Pottier, 1-23. London: Routledge.
——— 2002b. Globalizing indigenous knowledge. In *'Participating in development': approaches to indigenous knowledge* (eds) P. Sillitoe, A. Bicker & J. Pottier, 108-38. London: Routledge.
——— in press. The search for relevance: a brief history of applied anthropology. *History and Anthropology*.
———, J. BARR & M. ALAM 2004. Sandy-clay or clayey-sand? Mapping indigenous and scientific soil knowledge on the Bangladesh floodplains. In *Development and local knowledge: new approaches to issues in natural resources management, conservation and agriculture* (eds) A. Bicker, P. Sillitoe & J. Pottier, 174-201. London: Routledge.
——— & A. BICKER 2004. Introduction: hunting for theory, gathering ideology. In *Development and local knowledge: new approaches to issues in natural resources management, conservation and agriculture* (eds) A. Bicker, P. Sillitoe & J. Pottier, 1-18. London: Routledge.
SMITH, W. & T.C. MEREDITH 1999. Identifying biodiversity conservation priorities based on local values and abundance data. In *Cultural and spiritual values of biodiversity* (ed.) D.A. Posey, 372-5. London: Intermediate Technology Publications.
———, ——— & T. JOHNS 1996. Use and conservation of woody vegetation by the Batemi of Ngorongoro district, Tanzania. *Economic Botany* **50**, 290-9.
SPENCER, P. 2004. Keeping tradition in good repair: the evolution of indigenous knowledge and the dilemma of development among pastoralists. In *Development and local knowledge: new approaches to issues in natural resources management, conservation and agriculture* (eds) A. Bicker, P. Sillitoe & J. Pottier, 202-17. London: Routledge.
STAVENHAGEN, R. 1971. Decolonizing applied social sciences. *Human Organization* **30**, 333-44.
Survival International n.d. Website (*http://www.survival-international.org*).
SVARSTAD, H. & S.S. DHILLION (eds.) 2000. *Responding to bioprospecting: from biodiversity in the South to medicines in the North*. Oslo: Spartacus Forlag As.
TAX, S. 1975. Action anthropology. *Current Anthropology* **16**, 514-17.
Tebtebba Foundation n.d. Website (*http://www.tebtebba.org/tebtebba_files/wssd/indexa.html*).
TEKPAD n.d. Website (*http://ip.aaas.org/tekindex.nsf*).
VAN WILLIGEN, J. 1986. *Applied anthropology: an introduction*. South Hadley, Mass.: Bergin & Garvey.
WADDELL, E. 1972. *The mound builders: agricultural practices, environment, and society in the Central Highlands of New Guinea*. Seattle: University of Washington Press.
WALKER, D.H., F.L. SINCLAIR & G. KENDON 1995. A knowledge-based systems approach to agroforestry research and extension. *AI Applications* **9: 3**, 61-72.
WRIGHT, S. 1995. Anthropology: still the 'uncomfortable discipline'?. In *The future of anthropology: its relevance to the contemporary world* (eds) C. Shore & A. Ahmed, 65-93. London: Athlone Press.
WYNBERG, R. 2004. Rhetoric, realism and benefit sharing: use of traditional knowledge of Hoodia species in the development of an appetite suppressant. *Journal of World Intellectual Property* **7**, 851-76.
ZOBEL, G. 2004. Body fruits of the forest. *The Independent on Sunday Magazine*, 18 Apr., 41.

8

Meeting of minds: how do we share our appreciation of traditional environmental knowledge?

EUGENE HUNN

In this chapter I reflect on the relevance of ethnobiological knowledge to the conduct and writing of ethnography. I am in the midst of a major ethnographic writing project based on my recent Zapotec research. The monograph has grown such that the acquisitions editor at the academic press to whom I had made my initial submission advised me that it was of a size for two books rather than one. So I reorganized the material in two volumes, but they share a common title, *A Zapotec botany*, with contrasting subtitles (Hunn n.d.). While my title lacks the catchy phrasing one might hope for, it is at least descriptive of the content and modest in its claims. It is *A Zapotec botany*, for in fact there can be no single Zapotec botany, as there are several hundred Zapotec communities, each with its own autochthonous traditional knowledge specific to its traditional lands. I chose to title the book *A Zapotec botany* rather than the more customary *Zapotec ethnobotany* to emphasize my belief that the knowledge of plants widely shared by the people of San Juan Gbëë manifests a universal human scientific turn of mind. As we can see, the writing of ethnobiological ethnography begins with the title. But that is the least of the challenges ahead.

I will speak first about the difficulties we face communicating our findings to an audience that more often than not is woefully ignorant of the experiential reality of natural history that is one face of our subject matter. My suggestions in this regard are rather pedestrian but reflect both specific and general issues of cultural representation. The topic 'writing ethnobiology' called to mind Clifford and Marcus's influential volume *Writing culture* (1986). Thus I felt called upon to comment upon their critique of our ethnographic craft from my ethnobiological perspective.

Our task as ethnobiologists is not only faithfully to record what other cultures have learned in their daily encounters with their natural surroundings, but also to communicate an adequate appreciation of the perspicacity and intensity of the science of everyday life embodied in ethnobiological knowledge to an audience typically quite removed from the realities of extracting a livelihood with little more than head and hand.

When we speak of 'folk' biologies and of 'traditional' environmental knowledge, we self-consciously set our modern urban existence in opposition to a way of life that has endured far longer than ours and which I hope may continue, in some form or other, in the face of globalization. I see no reason to apologize for upholding this dichotomy nor for taking sides. I reject the notion that it is either romantic or patronizing to affirm the value of a way of life that has as its primary goal the continuation of that way of life through the generations – that is my definition of sustainability – and that is confined in space and through time to an intensely familiar landscape, one sufficient for the continued existence of an established community. Such confined communities are my definition of 'indigenous'.

This focus on 'traditional', 'indigenous', or, as some prefer, 'local' knowledge may be rationalized also by virtue of the fact that 'modern environmental knowledge' is of anthropological interest primarily because of what such bodies of knowledge lack. Experts aside, common knowledge of local biodiversity in modern societies can only be described as impoverished by contrast (Dougherty 1979), though not for any inability to perceive natural discontinuities (Boster 1987).

Language barriers

Granting that these folk sciences are worthy of our understanding, how do we get that point across to a wider public? The first task is to confront the language barriers, as I take it as given that human knowledge is first of all embodied in a particular human language. Language provides us with the tools for thought, for memory, and for imagination, and allows us to share with our fellows these thoughts, accounts of experience, and plans for meeting the future. The global conversation which is modern science, of course, depends on a myriad of special vocabularies hammered out through the communal practice of science. The Latinate names that have been imposed on the diversity of living things is a classic example, without which biology would be impossible. (In the future the language of genera and species might be replaced by a language that more directly replicates the genetic basis of biodiversity.

However, I doubt such a heroic reduction of the experiential reality of biological species and communities to nucleic acid sequences will ever be achieved.)

Scientific Latin is but one of thousands of human languages and reflects but one very special perspective on environmental diversity. What of those other thousands of human languages that have been put to the complementary task of grasping the essential details of the living environment? There is a curious paradox here. The dominant theoretical paradigm in contemporary linguistics implies that all human languages are but minor variations on a common theme of human understanding (Pinker 1994: 15-24). From this perspective it hardly matters which language we employ in our conversations about the world.

Yet cultural anthropology is inspired by a commitment to cultural relativism, in other words, that each cultural tradition defines a perspective on 'reality' that makes unique sense to those born and raised in that tradition. Sapir made the case emphatically: 'The worlds in which different societies live are distinct worlds, not merely the same world with different labels attached' (1929: 209). Sapir's argument has been referred to as the linguistic relativity hypothesis: 'Human beings do not live in the objective world alone ... but are very much at the mercy of the particular language which has become the medium of expression for their society ... [T]he "real world" is to a large extent unconsciously built up on the language habits of the group' (1929: 209).

Ethnobiological research has spoken on this issue, with authority. The relativists must grant that we all *do live in the same world*, with different names attached to the fundamental 'things'of that world, at least with respect to the organisms with which we share the planet. Biologists as well as anthropologists have noted the striking facility of intercultural communication at this basic descriptive level. Jared Diamond's essay on Fore ethno-ornithology is an early example (1966). Diamond had set up a field camp in the then rather mysterious central highlands of New Guinea in order to survey the little-known (to modern science) avifauna of this recently 'pacified' region of the world. He followed the example of his illustrious predecessor, Ernst Mayr, and hired local hunters to bring him a representative sample of the bird species of the local forests. The locals were willing, bringing specimens of each 'kind of bird' known to them. After careful comparison of these specimens with comprehensive collections of Papuan birds in museums of natural history, Diamond was able to recognize 120 Western ornithological species native to the area. These 120 species corresponded to 110 'species' named in their native language by the Fore. But just how, more precisely, did their 'species' correspond?

Fore recognize a life form *yakt*, roughly equivalent to English 'bird'. This is one of several *tábe aké*, or 'big names', that they distinguish. Such 'big names' include a variable number of *ámana aké*, or 'little names', which we may call 'species', though in Berlin's analytic vocabulary they are 'folk generics' (1992: 15-17). As noted, *yakt* included 110 *ámana aké*, which encompassed the 120 Western ornithological species Diamond recorded. Of these: ninety-three *ámana aké* corresponded in their denotative ranges 1:1 to the species recognized by academic ornithologists; eight *ámana aké* were over-differentiated, that is, they corresponded to just four of our species; while nine *ámana aké* were under-differentiated, encompassing twenty-two of our species. Not only is the agreement on the basic categories of the Fore world close if not perfect, the disagreements are readily explained. The eight cases of over-differentiation involve Fore naming male and female of four species of bird-of-paradise as in English we distinguish 'hens' from 'roosters'. The plumes of male birds-of-paradise are of extraordinary value to the Fore, in personal adornment and in exchange. Fore are, in any case, perfectly aware that their names refer to male and female respectively of four 'species'. Thus this apparent exception to the rule of intercultural concordance in fact affirms a more basic agreement, in recognizing the biological basis of the notion of species, as interbreeding populations. The cases of under-differentiation likewise make sense, given the Fore experiential reality. The species 'lumped' by the Fore were in each case closely related from the Western ornithological point of view: for example, the Fore *ámana aké atoku* refers to four species of frogmouths and owlet-nightjars (order: *Caprimulgiformes*). Furthermore, these four species are nocturnal and obscurely patterned birds. Thus, the Fore – who have a healthy respect for the spirits about at night – have limited opportunity to make their acquaintance. Nevertheless, the Fore recognize that there are distinct kinds of *atoku*, each with distinct vocalizations, though they do not recognize these 'species' distinctions nomenclaturally. In current ethnobiological analytic parlance such categories are 'covert folk species'.

Comparable examples might be multiplied *ad nauseum* from the ethnobiological ethnographic literature. At the same time it is clear that the correspondence of folk classification systems with that of modern evolutionary biology is in many intriguing details problematic. In part this is a function of the level of classification, that of Berlin's folk generic rank, demonstrating the least cultural flexibility, while life forms and other higher-order taxa often are deeply at odds with what modern science would like to accept as 'scientific', or, at least, phylogenetic. At the folk generic and specific levels of classification the contrasts with the generic and specific levels of the Linnaean system are largely with respect to over- and underdifferentiation, as in Diamond's Fore example, with folk systems typically more selective in according nomenclatural recognition to such natural kinds, particularly for small organisms or those of limited ecological or cultural salience (Hunn 1999). Such disagreements do not suggest any *fundamental* discord with respect to what is or is not a species. In fact, folk classifications at this basic descriptive level clearly recognize the biological basis of the species concept, namely that 'species' are self-reproducing populations of individual organisms, that is, they breed true, as the Fore treatment of bird-of-paradise demonstrates.

If our translation problems were limited to finding the corresponding scientific names for each local term, there would be no good reason to clutter our texts with native names, bedevilled by their orthographic idiosyncrasies, on the Chomskyan presumption – to a degree supported by ethnobiological research to date – that the underlying categories are innate and hence uninformative with regard to cultural peculiarities. However, the differences between a given folk classification and that which at the moment is judged most correct by modern science are highly informative of the range of possibility of the perception of order in nature. The classic debate with regard to just what *is* a cassowary is illustrative. Bulmer's classic essay on this detail of the ethno-ornithology of the Kalam of Papua New Guinea – neighbours of the Fore of Diamond's study – 'Why is the cassowary not a bird?' (1967) explores the complex role that cassowaries play in the lives of the Kalam. This is *not* a question primarily of the ability of Papua New Guinea highlanders to perceive 'true' taxonomic relationships, since what is in fact a 'bird' is a less than obvious question, particularly in light of the recent dinosaur problematic (e.g. Gauthier & Gall 2001). Both academic ornithologists and Kalam observers agree that cassowaries are quite unlike the great majority of 'birds' (Fig. 1), yet for Kalam these idiosyncrasies express more profound differences, as the cassowary is in some sense more human than bird. In this regard Kalam understandings are based on radically different premises than our own.

What's in a name?

A careful consideration of the morphology and semantics of names for plants and animals offers rewards beyond simply identifying points of agreement or disagreement with modern science. The 'descriptive force' of native plant and animal names is revealing as well of what is *seen* most clearly by Native eyes. For example, *binomial* names, composed of a 'head' which in isolation names a superordinate plant or animal category (which itself may be binomial) plus a modifying attributive, explicitly mark taxonomic relationships, while the attributive may signal as well some distinctive quality of the category useful in distinguishing it from its close allies. Such systematic naming patterns are by no means restricted to Linnaean classification: for example, 'black oak' is a kind of 'oak' distinguished from, say, 'white oaks' and 'red oaks' by the blackish bark of mature trees, while Tzeltal Maya differentiate several local species of 'robins', naming them binomially as 'spiny robin', 'yellow robin', 'black robin', 'red robin', and so on (Fig. 2). Note, however, that binomials may name folk generics, in

Figure 1. *Kobty, Casuarius bennetti,* Dwarf Cassowary, drawing by Chris Healey (Majnep & Bulmer 1977: plate R, reproduced with the permission of Chris Healey).

which case they are 'productive primary lexemes', or folk specifics, in which case they are 'secondary lexemes', according to Berlin (1992: 26-31). Care is required, however, to distinguish true binomials from pseudobinomials, in which case the head element names a category which could be but which is not in fact superordinate to the taxon so named, for example 'silverfish', which

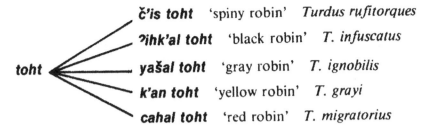

Figure 2. Tzeltal robins (Hunn 1977: 56).

is silver alright but not a fish, though it bears some superficial resemblance to one.

Other names are onomatopoetic, characteristic of many bird and amphibian names (Berlin & O'Neill 1981), while still others indicate features distinctive, but never definitive, of the category named, marking, for example, colour, size, shape, taste, odour, behaviour, habitat preference, and/or cultural utility. Thus local names, carefully analysed, are rich in cultural nuance and highly informative of contrasting cultural perspectives on the living environment.

Practical implications

Thus an ethnobiological ethnography should at least summarize local vernacular names and illustrate in some detail how those names are applied to the local flora and fauna, which, to the extent it is an exotic flora and fauna not familiar to the majority of readers, can only be accurately characterized in the Latinate medium of the Linnaean 'etic grid'. Literary critics might dismiss this injunction as a rationalization that obscures a more telling motive, to wit, that by citing scientific and native names we validate the authenticity of and give authoritative weight to our accounts. However, this is at best but a small part of the story. More to the point is the fact that an adequate appreciation of the subtlety of local understandings of natural history is hampered by failing to appreciate the limitations of English for translating native vocabulary. For example, the definitive ethnographic account of the Sanpoil Indians of the Columbia Plateau of northwestern North America described local harvests of five species of *Camassia*, the Latin name for the camas lily, a key local staple (Hunn 1990; Ray 1933). Students of Plateau ethnography searched in vain for these species, as just one species was known to occur in the region. The ethnographer assumed that the local English vernacular term 'camas', used by his bilingual Indian consultants, was equivalent to the Latin genus *Camassia*, a term originally borrowed from a different Indian language to refer to this native lily. In the English vernacular, however, the term was generalized to refer to a variety of edible roots and bulbs of distinct botanical families. The 'camas' species in question proved to be not only camas, properly so called, but also several species of *Lomatium*, a genus in the carrot family. To keep faith with the indigenous perspective, we can avoid neither the indigenous nor the Latin linguistic medium. As a consequence, we belabour our ethnographic stories with hundreds of italicized terms that carry no ring of familiarity for the average reader, as in the example below from *A Zapotec botany*:

> One may 'lift' *aires* bedeviling one's house – which inflict nightmares (***mcàal***) and insomnia induced by a 'fright' – with *limpias* 'cleanings' or 'sweepings', either by burning specific curative plants – as for example, balsam fir (***yàg-lgâzh***, *Abies guatemalensis*) – in the house or literally sweeping (***rliòob***) the room (or the patient) with the plant. One such plant is ***yàg-ngùd-guèy-pcàal*** 'nightmare white zapote' (*Bysonima crassifolia*). Another is ***yàg-bdìin*** 'bad airs tree', a.k.a. ***guìzh-zhwèe*** 'injury herb' (*Eupatorium mairetianum*). Leaves of these specially-designated herbal remedies are mixed with leaves of rue (*Ruta graveolens*) and white zapote (***yàg-ngùd-guèy***, *Casimiroa edulis*) to prepare an infusion for bathing or to be taken as a tea (Hunn n.d.: part 1, 241).

How may we avoid alienating the audience for our ethnographic stories so burdened?

Narrative layers

One solution might be to write our accounts in layers, each designed to appeal to a distinct audience. I will describe three such narrative layers: the master narrative, the technical narrative, and the monographic narrative. The top layer would address the most generalized audience and be designed first of all to convey the power and complexity of our ethnobiological material to an audience with limited firsthand knowledge or appreciation of natural history. Such narratives must capture the reader's attention. Distraction can be deadly. Gary Nabhan (1982) and Richard King Nelson (1983) have shown us the way, as have authors not known first of all as ethnobiologists, but who are gifted interpreters of natural history, for example Barry Lopez (1986) and Peter Matthiessen (1978), whose appreciation of nature is sufficiently nuanced to encompass the human. Let me quote one brief example:

> My pickup truck bounced along over the washboard road. Amadeo, Remedio, and I pointed out plants to each other as we went – the bristle-topped senita cactus, heavy-trunked ironwood trees, and odorous, yellowish-green croton shrubs.
> We edged over a rise, and all of a sudden the desert was whisked away – palms and cottonwoods reached above the horizon, and teal splashed up into the air. Amadeo grabbed his field glasses – a white-faced ibis down on the mudflats of the pond, and a couple of pigs foraging in the saltgrass.
> Remedio sighed, knowing that this place was the place he had heard of: '*Ki:towak*' (Nabhan 1982: 90-1).

We may call these *master narratives*. Master narratives are designed to *convince* the reader with all the rhetorical powers at our command, to convince intellectually, emotionally, and sensually. Master narratives are a form of propaganda, but heartfelt. Our master narratives may argue that other peoples at other times and places deserve our respect for their humanity, their intellect, their curiosity, their sense of place and attachment to family and community.

This may be the master narrative of anthropology, which takes specific ethnobiological forms. For example, in our analyses of folk biological classification we intend to show how folk biology is a science, comparable to modern biological science, if judged on its own terms. Likewise, we may analyse traditional resource management practices to demonstrate their sustainability or the comparability of animist moral narratives to the biophilia of E.O. Wilson (1984). However, competing master narratives may spin the evidence in quite another direction, as when a neo-Darwinian master narrative defines our common humanity by affirming, *ceteris paribus*, that short-term self-interest shall prevail (Hunn, Johnson, Russell & Thornton 2003) or when a conservation biology master narrative portrays humanity as a lethal virus infecting the biosphere.

If we consider ourselves scientists – as I believe most ethnobiologists do – we will not be satisfied with master narratives. Rather, we will seek to convince our colleagues not solely by rhetoric, but also through the soundness of our analyses and by appeals to evidence and logic. This brings us to the second ethnographic layer, our *technical narrative*. As narratives, these texts are argumentative still, but must address the issues in more detail, for example citing the academic literature to highlight the status of particular debates that engage various scholarly communities. For instance, the technical narrative beneath our master narrative highlighting nomenclatural curiosities or the impressive detail of native taxonomic distinctions might elaborate on the fit of the particular case to general principles, such as those of Berlin. Or, if one elaborates a master narrative to the effect that the local herbal apothecary demonstrates a sophisticated grasp of the pharmacological properties of local plants, the underlying technical narrative might include comprehensive lists of medicinal plants with detailed recipes and notes to the pharmacological literature by way of measuring the empirical validity of local herbal remedies:

Adiantum poiretti (Sahagun 1950-1969: book 11, 196; Hernández 1942-1946: vol. 2, 340). Tannins are hemostatic. The plant is used as a diuretic in Guatemala (Orellana 1987: 174) and by the Iroquois (Moerman 1986: 11, 19). Diuretic action satisfies Aztec criteria for a fever remedy. *A. lucida* is used in the West Indies as a febrifuge (Ayensu 1981: 152) (Ortiz de Montellano 1990: 254).

One might finally elaborate a third layer, that of the *monographic narrative*, that would explore the limits of the data, acknowledging ambiguities and varied opinions among consultants, summarizing the consultant sample with respect to how representative it might be of the wider community, and detailing taxonomic uncertainties in voucher specimen determinations. An extreme case might be Laughlin's characterization of the apparently chaotic naming of macrofungi by his Zinacantecan consultants (Fig. 3). These two or three levels of narrative might be managed in print, for example, in banks of footnotes printed in progressively smaller fonts to conserve space, available for ready consultation, but only if the reader feels compelled to break off from the master narrative for a technical detour. Alternatively, supplementary exemplification, documentation, and/or referencing might be relegated to appendices or marginal text boxes. Of course, if publication were digital, these layers could be seamlessly integrated.

A system of narrative layering might go some way towards addressing the fact that the audiences for ethnobiological ethnography are highly diverse. The audience for our master narratives about people, plants, and animals is ample and the potential audience – that is, those who *should* confront our master narratives – larger still. However, the audience for our more detailed technical and monographic narratives is quite limited. In part this is a consequence of what makes our master narratives so broadly appealing, the inextricably multidisciplinary nature of our subject matter, encompassing as it does cultural anthropology, archaeology, linguistics, cognitive psychology, systematic and evolutionary biology, auto- and syn-ecology, pharmacology, and agronomy, not to mention epistemology and ethics.

At the level of our master narrative these various disciplinary perspectives combine to enhance our audience, but at the level of our technical narrative they *intersect* to exclude all but the most dedicated interdisciplinarian, restricting at the same time our publication options. By writing two or three books in one, targeting general and specialists audiences, we may perhaps overcome this paradoxical limitation.

Limitations of the written word for ethnobiological narratives

Though we may address the problem of writing for our diverse audiences, this does not address a more fundamental problem, inherent in *writing* itself. Words simply cannot convey to the uninitiated the experience of biodiversity, the complex patterns and shading of form, colour, scent, and sound of living things at home in their native habitat. Children of the societies we describe do not learn to recognize the hundreds of plants and animals they come to know well by their tenth birthdays by means of verbal descriptions, any more than they learn them by means of dichotomous keys. Rather, children are programmed

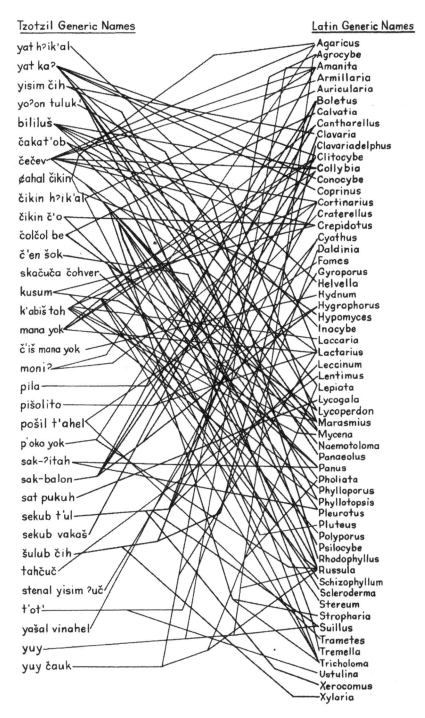

Figure 3. Graphic representation of the correspondence between Tzotzil and scientific names based on Laughlin's raw data (1975: 8, reproduced with permission).

to learn to recognize, visually but also in all their sensory modalities, family resemblances among living kinds and to associate names in their native language with these patterns (Gelman & Coley 1991; Hunn 2002). And they learn in powerful social contexts, from their parents, elders, and peers, though typically without formal instruction.

I believe that an effective ethnobiological ethnography should replicate to the extent possible this *natural* learning process, given that our audience will most certainly lack the experiential basis to appreciate that which our native consultants implicitly understand. In short, names are not enough. We must *illustrate* our ethnographies, ideally illustrating the full complement of species present in full colour, while highlighting the rich contextual meanings of each.

I recognize the valiant efforts of my colleagues to achieve some semblance of this goal, yet I would have to judge our efforts to date lacking on this account. To cite just a few representative examples, Takashi Ijichi's evocative water colours in Rea's *At the desert's green edge* (1997) (Fig. 4), Terry Bell's fine pen-and-ink drawings that illustrate Breedlove and Laughlin's monumental *The flowering of man* (2000) (Fig. 5), and Christopher Healey's birds in Majnep and Bulmer's collaboration, *Birds of my Kalam country* (1977) (Fig. 6). However, more often than not we have had to make do with the occasional illustrative example or stark black-and-white photographic images of dead plants and animals pulled from their museum shelves (e.g. Berlin, Breedlove & Raven 1974; Hunn 1977) (Fig. 7) or the mind-numbing sameness of pure text.

I have recently hit on one rather effective, if partial, solution, at least for the botanical images: scans. Scanners are now quite portable and provide excellent resolution, though in a narrow focal plane and limited to the 9 × 12 inch frame of the glass. (I scan fresh plants at 300 pixels/inch; each scan is several hundred kilobytes to a megabyte if saved as a jpeg file [Fig. 8].) With the possibility of publishing print books with CD supplements or on-line, the cost of reproducing a large number of colour illustrations is no longer prohibitive. We may even incorporate sound clips of birds singing and frogs croaking. We may illustrate onomatopoeia by directly comparing the spoken name with the animal's vocalizations (and with our practical orthographies). I believe such multimedia ethnographies soon will be standard practice in ethnobiology.

The critics

Writing culture edited by Clifford and Marcus (1986) challenged ethnography to abandon the scientific pretence of a search for truth and to accept ethnography as literature. This requires not only that our writing be judged by the aesthetic canons of literature, but also that we bring to our writing a

ñiádam chu'igam, ge'egeḍ haahagam, s-tadañ haahag *Malva parviflora*

Figure 4. Takashi Ijichi sumi-e illustration of *Malva parviflora* (Rea 1997: 235, reproduced with the permission of Takashi Ijichi).

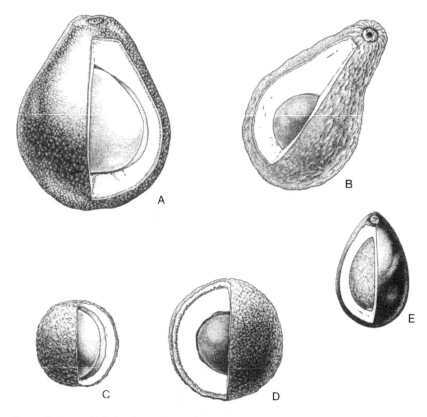

Figure 5. Pen-and-ink drawings of avocado cultivars by Terry Bell (Breedlove & Laughlin 2000: Fig. 30, reproduced with permission).

sophisticated self-consciousness of narrative devices and the power they wield, to subject our writing to the critical scrutiny of contemporary literary deconstruction.

Those of us who were writing culture before Clifford and Marcus may dismiss them as upstart literati. Yet the issues raised by the 'literary turn' that they helped foment in anthropology are not entirely irrelevant to our craft, at least at the level of our master narratives. However, I would argue that our technical narratives should be responsive to quite different, non-literary standards, to wit, those of scientific reportage, and in particular our monographic texts are best considered *anti-narratives*. That is, the goal is not to convince by rhetoric or appeals to the best of our evidence, but, rather, self-consciously to expose our arguments to their foundations, however shaky these may be,

Figure 6. Chris Healey's lesser bird-of-paradise (Majnep & Bulmer 1977: 137, reproduced with the permission of Chris Healey).

Figure 7. Photograph of voucher specimen of *Lantana hirta* (Berlin, Breedlove & Raven 1974: Fig. 8.6, reproduced with permission).

playing the devil's advocate. This is ultimately still an effort to convince, but by pre-empting all reasonable alternative interpretations. Postmodern literary critics may deny that this is possible (Clifford 1986: 3).

Clifford and Marcus positioned themselves in the avant-garde, but their critical ruminations came rather late for ethnobiologists, as we were well ahead of the curve they traced. Consider Clifford's call for dialogic ethnographic narratives that allow the 'subaltern' voice of the Other to be heard (Clifford 1986: 17). I believe still the finest example of such an effort in ethnographic writing is Majnep and Bulmer's *Birds of my Kalam country*, first published in 1977, nine years before Clifford and Marcus's volume. Bulmer, the ethnobiologist, and Majnep, the native consultant, collaborated on this book in the most profound sense, though Bulmer was the titled professor, while Majnep had had no formal post-primary schooling. Majnep's text appears in full, translated by Bulmer, then set beside Bulmer's complementary account, which brings to bear his modern scientific perspective to complement Majnep's richly experiential and culturally informed account. Such collaboration is perhaps an essential quality of ethnobiological research. Our research builds on conversations among *aficionados* of natural history. We meet our indigenous colleagues on the common ground of our shared fascination with the natural world.

Figure 8. Scanned image of *Solanum lanceolatum*.

We do not need Clifford and Marcus to point us in the direction of truly collaborative ethnography or to instruct us in the fine art of narrative writing, as we count among us true masters of the art. Nor do we need Clifford and Marcus to inform us that our 'earlier modes of unselfconscious representation' are incoherent (Rabinow 1986: 250), that our theoretical projects are 'in disarray' (Marcus 1986: 263), or that our discipline is 'in crisis' (Clifford 1986: 3). Clifford and Marcus's critique of ethnographic writing was raised on the philosophical quicksands of postmodernism, rejecting classical epistemology in favour of a hermeneutical game of 'knowledge without foundations' (Rabinow 1986: 236). Ethnobiology plays a different 'game', I believe, because we confront in our ethnobiological practice powerful evidence that ethnobiological knowledge is firmly grounded in concrete encounters of human minds with the natural world. Marcus grudgingly recognizes this contrast between the reflexive subtlety but substantial poverty of Stephen Tyler's collaborative ethnographic writing (1988) and the straightforward empirical emphasis of Bulmer's collaboration with Ian Saem Majnep. Bulmer has the last word in putting aside the ethical agonizing so prominent in postmodern critical analyses. He muses, 'Surely it is better to go on doing something we [that is, Majnep and himself] both enjoy so much, than to be paralyzed into inactivity by ethical anguish' (quoted in Marcus 1991: 41).

Conclusion

Ethnobiological ethnography poses particular challenges due to the eclectic diversity of the subject matter treated and the doubly alien nature of the substantive material to be described – involving alien cultural perspectives expressed in an alien language and an unfamiliar flora and fauna. I argue that the alien cultural perspective requires extensive reference to technical nomenclature in both the language of the alien culture and the, for many, equally alien language of modern science. I suggest that one solution to these peculiar obstacles to effective communication to our target audiences is to compose layered narratives: (1) a master narrative designed to convince a broad lay audience, (2) a technical narrative that marshals factual evidence and academic argument to convince a potentially sceptical scholarly audience, and finally (3) a monographic narrative that exposes the weaknesses along with the strengths of the evidentiary basis of the technical narrative. Integrating these layers effectively will require creative design choices and sophisticated multimedia technology. Ethnobiological ethnography should not be judged primarily as literature but rather with regard to how effectively we meet the expectations of the diverse audiences addressed at each narrative layer.

REFERENCES

BERLIN, B. 1992. *Ethnobiological classification: principles of categorization of plants and animals in traditional societies*. Princeton: University Press.

———, D.E. BREEDLOVE & P.H. RAVEN 1974. *Principles of Tzeltal plant classification: an introduction to the botanical ethnography of a Mayan-speaking community of highland Chiapas*. New York: Academic Press.

——— & J. O'NEILL 1981. The pervasiveness of onomatopoeia in the Jivaroan language family. *Journal of Ethnobiology* 1, 95-108.

BOSTER, J. 1987. Agreement between biological classification systems is not dependent on cultural transmission. *American Anthropologist* 89, 914-19.

BREEDLOVE, D.E. & R.M. LAUGHLIN 2000. *The flowering of man: a Tzotzil botany of Zanacantán* (abridged edition). Washington, D.C.: Smithsonian Institution.

BULMER, R.N.H. 1967. Why is the cassowary not a bird? *Man* (N.S.) 2, 5-25.

CLIFFORD, J. 1986. Introduction: partial truths. In *Writing culture* (eds) J. Clifford & G.E. Marcus, 1-26. Berkeley: University of California Press.

——— & G.E. MARCUS (eds) 1986. *Writing culture: the poetics and politics of ethnography*. Berkeley: University of California Press.

DIAMOND, J.M. 1966. Zoological classification system of a primitive people [the Fore of Highland Papua New Guinea]. *Science* 131, 1102-4.

DOUGHERTY, J. 1979. Learning names for plants and plants for names. *Anthropological Linguistics* 21, 298-315.

GAUTHIER, J. & L.F. GALL (eds) 2001. *New perspectives on the origin and early evolution of birds: proceedings of the International Symposium in honor of John H. Ostrom, February 13-14, 1999, New Haven, Connecticut*. New Haven: Peabody Museum of Natural History Yale University.

GELMAN, S.A. & J.D. COLEY 1991. Language and categorization: the acquisition of natural kind terms. In *Perspectives on language and thought* (eds) S.A. Gelman & J. Byrnes, 146-96. New York: Cambridge University Press.

HUNN, E.S. 1977. *Tzeltal ethnozoology: the classification of discontinuities in nature*. New York: Academic Press.

——— 1990. *Nch'i-Wána 'The Big River': mid-Columbia Indians and their land* (with James Selam and family). Seattle: University of Washington Press.

——— 1999. Size as limiting the recognition of biodiversity in folk biological classifications: one of four factors governing the cultural recognition of biological taxa. In *Folkbiology* (eds) D.L. Medin & S. Atran, 47-69. Cambridge, Mass.: Harvard University Press.

——— 2002. Evidence for the precocious acquisition of plant knowledge by Zapotec children. In *Ethnobiology and biocultural diversity* (eds) J.R. Stepp, F.S. Wyndham & R.K. Zarger, 604-13. Athens: University of Georgia Press.

——— n.d. *A Zapotec botany*, parts 1 and 2. Manuscript under review.

———, D. JOHNSON, P. RUSSELL & T.F. THORNTON 2003. Huna Tlingit traditional environmental knowledge and the management of a 'wilderness' park. *Current Anthropology* 44, 79-104.

LAUGHLIN, R.M. 1975. *The great Tzotzil dictionary of San Lorenzo Zinacantan*. (Smithsonian Contributions to Anthropology 19). Washington, D.C.: Smithsonian Institution Press.

LOPEZ, B. 1986. *Arctic dreams: imagination and desire in an Arctic landscape*. New York: Scribner.

MAJNEP, I.S. & R.N.H. BULMER 1977. *Birds of my Kalam country*. Oxford: University Press.

MARCUS, G.E. 1986. Afterword: ethnographic writing and anthropological careers. In *Writing culture* (eds) J. Clifford & G.E. Marcus, 262-6. Berkeley: University of California Press.

——— 1991. Notes and quotes concerning the further collaboration of Ian Saem Majnep and Ralph Bulmer: Saem becomes a writer. In *Man and a half: essays in Pacific anthropology and ethnobiology in honour of Ralph Bulmer* (ed.) A. Pawley, 37-45. Auckland: The Polynesian Society.

MATTHIESSEN, P. 1978. *The snow leopard.* New York: Viking Press.

NABHAN, G. 1982. *The desert smells like rain: a naturalist in Papago Indian Country.* New York: North Point Press.

NELSON, R.K. 1983. *Make prayers to the raven: a Koyukon view of the Northern Forest.* Chicago: University Press.

ORTIZ DE MONTELLANO, B.R. 1990. *Aztec medicine, health, and nutrition.* New Brunswick: Rutgers University Press.

PINKER, S. 1994. *The language instinct: how the mind creates language.* New York: W. Morrow & Co.

RABINOW, P. 1986. Representations are social facts: modernity and post-modernity in anthropology. In *Writing culture* (eds) J. Clifford & G.E. Marcus, 234-61. Berkeley: University of California Press.

RAY, V.F. 1933. The Sanpoil and Nespelem: Salishan peoples of northeastern Washington. *University of Washington Publications in Anthropology* 5.

REA, A. 1997. *At the desert's green edge: an ethnobotany of the Gila River Pima.* Tucson: University of Arizona Press.

SAPIR, E. 1929. The status of linguistics as a science. *Language* 5, 207-14.

TYLER, S.A. 1988. Words for deeds and the doctrine of the secret world: testimony to a chance encounter somewhere in the Indian jungle. In *The unspeakable: discourse, dialogue, and rhetoric in the postmodern world,* S.A. Tyler, 66-88. Madison: University of Wisconsin Press.

WILSON, E.O. 1984. *Biophilia.* Cambridge, Mass.: Harvard University Press.

Index

Ackerknecht, E.H., 119
Adams, C., 127, 137
adornment studies, 124
agricultural regression model, 102-7
Aguilar, L., 129
Amazonia
    and agricultural regression model, 102-7
    anthropogenic forest formations, 99-102
    historical ecologies, 97-115
    landscape history, 111-12
    terra preta do Indio (black soil), 161-2
    trekkers and foragers, 107-11
Anderson, E.N., 111
Anderson, G.J., 129
Anderson, R., 122
Ankli, A., 135-6
anthropogenic forest formations, in Amazonia, 99-102
anthropomorphism, 63, 69
anti-narratives, 190-2
applied anthropology, 147-76
applied ethnobiology, 14-15
archaeological evidence, 17-18
    for human evolution, 56-9
    for natural history intelligence, 63-6
    for subsistence in tropics, 77-95

artemisinin, 166
Aspilia spp., 132
Atran, S., 64

Balám, G., 122-3
Balée, W., 18, 98, 100-1, 102-3, 104-5, 110
banana cultivation, 81-2
Bandoni, A.L., 128
Bangladesh, 153-6
Bartlett, H., 150
Barton, H., 83
Bastien, J., 124-5, 126
Basualdo, I., 128
Bennett, J.W., 147
Berlin, B., 8-9, 10, 13, 65, 118, 133, 150-1, 181-2
Berlin, E.A., 13, 118, 123, 127, 137
Berreman, G.D., 151
binomial names, 181-2
biocultural medical anthropology, 117, 122
biocultural synthesis, 15-18
biodiversity conservation, 156-9
Bioharmony, 165-6
biomedicine, 120, 122-3
bio-prospecting, 164, 165-6
bird names, 39-44, 179-80, 181
Bloch, M., 11
Blurton-Jones, N., 63

body
  body-part nomenclature, 124-5, 127
  concepts of, 123-7
  physical and social, 123-4
botanical pharmacopoeias, 133-4
Boulos, L., 128
Bourdieu, P., 124
Bowdery, D., 82
brain size, 56, 62
Breedlove, D.E., 188
Brett, J.A., 135
Brown, C.H., 124
Browner, C., 122, 130, 134-5
Buckley, A.D., 126
Bulmer, R.N.H., 8, 11, 181, 188, 192, 194
Burkill, I.H., 84

Caceres, A., 129
Camassia (camas lily), 183-4
Cano, O. 129
Carey, S., 64
Carroll, L., 29
Carruthers, P., 69
cassowary, 181
Castro, E. 127, 137
category formation, 9-10
Cavalli-Sforza, L.L., 12
cha'lam tsots (ailment), 123
chemical ecology, 130-3
chimpanzees, 56, 59, 60, 132-3
Chinemana, F., 128
chiropractors, 122
Clement, C., 101-2, 111
Clifford, J., 8, 177-8, 188, 190, 192, 194
Coe, F.G., 129
co-evolutionary paradigm, 15-18
cognition and culture, 8-10
cognitive fluidity, 69
collaborative ethnography, 192-4
Conklin, H.C., 7, 8-9, 10, 150-1
conservation management, 156-9
Coussio, J.D., 128
critical medical anthropology, 117
Crumley, C., 98, 111
cultural diffusion, 159-64
cultural diversity, and genetic similarity, 59
cultural relativism, 179

culture-bound syndromes, 123
cupuacu (*Theobroma grandiflorum*), 166

Davis, B.L., 47
Denevan, W., 101, 105
d'Errico, F., 58
development, 147-76
  alternative, 167-70
  and commercial use of knowledge, 164-7
  ethnobiology and indigenous knowledge, 151-2
  facilitation of local solutions, 156-9
  furthering cultural diffusion, 159-64
  introduction of exogenous technology, 152-6
  sustainable, 168-70
de Zoysa, I., 128
Diamond, J.M., 179-80
Donald, M., 47-8
Douglas, M., 9, 123
Drummond, R.B., 128
Dunbar, R.I.M., 61

ecological brain, 61-3
economic botany, 14
Eisenberg, L., 122
Ellen, R.F., 151
ethnobiological classification, 8-9, 29, 60-1, 65-6
ethnobiological ethnography, 177-96
  and critics, 188-94
  illustrations, 188
  language barriers, 178-81
  limitations of written word, 186-8
  names, 181-3
  narrative layers, 184-6
  practical implications, 183-4
ethnobiology
  applied, 14-15
  and applied anthropology, 147-76
  centrality to anthropology, 18-20
  contributions to anthropology, 3-4
  defined, 3
  first phase/dimension, 2
  historical background, 1-4, 148-51
  second phase/dimension, 2
ethnobotanical inventories, 128-9
ethnoecology, 14

ethnographic databases, 160
ethnography as literature, 188-90
ethnomedicine, 117-45
ethnophysiology, 123-7
ethnosemantic methods, 6
ethnozoological nomenclature, 29-54
Etkin, N.L., 13, 129
Evans-Pritchard, E.E., 151
evolution of human mind, 55-75

Fabrega, H., 125
farming, origins of, 58
Fausto, C., 108
Feld, S., 7
fish knowledge, local, 154-5
fish names, 38
folk classification, 10, 179-81
folk healers, 117-18, 120, 121-2
food-processing technology, 131
foragers, 84-90, 102-11
Ford, R.I., 1, 18, 19
Fore, 179-80
forest formations in Amazonia, anthropogenic, 99-102
fossil record for human evolution, 56-9
Foster, G.M., 119
Fowler, D.D., 111
FOXP$_2$ gene, 68
Frake, C.O., 13, 150-1

game parks, 157-9
García, H., 122-3
genetic similarity, and cultural diversity, 59
Getui, M., 158
Gollin, L., 136
Golson, J., 78
Goodale, J.C., 65
Green, E.C., 118

habitus, 124
Hardesty, D.L., 111
Harman, R.C., 120
Harris, D.R., 17-18, 101
Hather, J.G., 80, 81
Haudricourt, A., 2
health care systems, 120
Heinrich, M., 134, 135-6
herbal treatments, 117-45

Hernández Cano, J., 128
Hill, K., 104
Hillman, G.C., 90
Hinton, L., 31, 37
historical ecology, 17, 18
 and Amazonia, 97-115
 origins and principles, 98-9
'Hmmmmm' (communication system), 67-9
Holland, W.R., 125
hominins, 56-70
Homo antecessor, 57
Homo erectus, 57, 58
Homo ergaster, 56-7, 61, 64
Homo floresiensis, 56, 58
Homo habilis, 56-7
Homo heidelbergensis, 57, 66
Homo helmei, 67-8
Homo neanderthalensis, 57, 58, 66, 67
Homo rudolfensis, 56-7
Homo sapiens, 9, 57-8, 59, 64, 69
Hoodia sp., 165
Howard, P., 12
Huaorani, 108-10
Hubbard, E.M., 48
Huffman, M.A., 132-3
human chemical ecology, 130-3
human mind, evolution of, 55-75
Humphrey, N., 60
Hunn, E., 5, 7, 8, 184
Hutchings, A., 129

illustrations, ethnographic, 188
indigenous knowledge
 and applied anthropology, 147-76
 and development, 147-76
 and ethnobiological ethnography, 178
Indigenous Peoples' Organizations, 168
Ingold, T., 15-16
intellectual property rights, 164-7
intentionality, orders of, 61-2
international development *see* development
interpretative medical anthropology, 117
intuitive biology, 62-6, 70

Jackson, F.L.C., 131
Jakobson, R., 31-2
Jamaica, 126

Jara, V., 123
Jesperson, O., 34
Johns, T., 130-1, 132
Jones, G.E.M., 90
Journet, N., 108

Kalam, 181
Kari-Oca Declaration, 168
Keen, S.L., 131
Keil, F., 64
Kennedy, J.S., 63
Kent, S., 108
Kern, V., 111-12
Kiefer, D., 133
Kimberley Declaration, 169-70
Kirby, S., 68
Kleinman, A., 120, 122
knowledge transmission/acquisition
  and hunter-gatherers, 64-5
  and social organization, 10-12
Köhler, W., 34
Konner, M.J., 63
Kubo, I., 131
Kurtz, W.B., 120

landscape history, 111-12
language
  and ethnobiological ethnography, 177-96
  and ethnozoological nomenclature, 29-54
  origins of, 66-9, 70
  and translation of knowledge systems, 6-8
Latin, 5-6, 178-9
Latin American humoral system, 125, 134
Laughlin, R., 127, 186, 188
Leach, E., 9, 99
Leonti, M., 134, 136
Lévi-Strauss, C., 2
Lewis, G., 13
linguistic relativity hypothesis, 179
Linnean grid, 4-6
Logan, K., 121
Luber, G.E., 123

Macgillivray, J., 89
MacNeilage, P., 47

Maffi, L., 103, 127
magical medicine, 119
Majnep, I.S., 8, 188, 192, 194
mangroves, 89-90
manioc cultivation, 80-1, 83
Marcus, G.E., 8, 177-8, 188, 190, 192, 194
Martin, G.J., 149
master narratives, 8, 184-5, 186
Mauss, M., 124
Mavi, S., 128
Maya, 120, 125-6, 127, 135-6, 137-8
medical anthropology
  history of ethnomedical approach, 118-23
  and medical ethnobiology, 12-14, 117-45
medical ecology, 117, 121
medical ethnobiology, 12-14, 117-45
medical ethnobotany, 127-37
Meggers, B., 99
Mendiondo, M.E., 128
mental modularity, 59-61, 62
Meredith, T.C., 158
Metzger, D., 125, 151
me' winik (ailment), 123
Mithen, S., 9, 48
mobility, 107-8
Moerman, D.E., 128, 133
monographic narratives, 186, 190-2
Moore, D.R., 89
Morgan, L.H., 10
multimedia ethnographies, 188
music, origins of, 67, 68-9

Nabhan, G., 184
narrative layers, 184-6
Nash, J., 125
natural history intelligence, 62-6
naturalistic disease causation, 119, 121-2
natural selection, 59-60
Neanderthals, 57, 58, 66, 67
New Ethnography, 150
New Guinea, 78, 162-3, 179-80, 181
Newton, P.N., 132
Nichols, J., 31, 37
Nishida, T., 132

Ohala, J., 31, 37, 45-6
optimal foraging theory, 104

Ortiz, M., 128
Ortiz de Montellano, B.R., 122, 129-30, 185

Paget, R., 46-7
palaeoanthropology, and ethnobiology, 55-6
parenchyma analysis, 80-1
participatory methodologies, 158-9
Paz, V., 83
Pearsall, D.M., 81
Pemberton, R., 133
Perry, L., 83, 90
personalistic disease causation, 119, 120, 121
phonaesthesia, 31-49
  and consonants, 45
  early studies, 31-4
  in ethnozoological nomenclature, 38-42
  experiments on, 34-7
  perceptual properties, 37
  and size/shape/movement, 37-44, 49
  and vocal mimesis, 46-9
phytolith analysis, 81-2
Piperno, D.R., 81, 83
plant medicines, 127-37
  and commercial use of knowledge, 165-6
  documentation and evaluation, 128-30
  efficacy, 129-30
  evolutionary basis, 130-3
  and human chemical ecology, 130-3
  organoleptic characteristics, 135-7
  selection criteria, 133-7
Popoluca, 134, 136
Posey, D., 8, 100, 101
proto-language, 66-9
pseudobinomial names, 182-3

qinghao (plant), 166
Qollahuaya, 126
Queensland almond (*Elaeocarpus bancroftii*), 87-8
Queensland rain forest, pre-European subsistence in, 84-8

Raffauf, R.F., 129
Ramachandran, S.V., 48

Ramirez, F., 134
Rea, A., 188
Rindos, D., 101
Rival, L., 17-18
Rondina, R.V., 128
Roosevelt, A., 101
root and tuber cultivation in tropics, 79-84
  parenchyma analysis, 80-1
  phytolith analysis, 81-2
  starch-grain analysis, 82-4
Rubel, A.J., 122

Samayoa, B., 129
San people, 165
Sapir, E., 31, 33-4, 103, 179
Sauer, C., 111-12
scanned images, 188
Schultes, R.E., 8, 129
Scotch, N., 118
segmentation, and language, 68-9
Seifu, M., 132-3
Shepard, G.H., 136-7
sickness in humans, 117-45
Sierra, A., 122-3
Sillitoe, P., 4
Silver, D., 125
size-sound symbolism, 34
Smith, W., 158
social body, 123-4
social brain hypothesis, 61-3
social intelligence, 61-3
social organization, and knowledge transmission, 10-12
soils
  and cultural diffusion, 161-4
  local knowledge of, 153-4, 155-6
  soil mounds, 163-4
species concept, 179-81
starch-grain analysis, 82-4
Steward, J., 98-9, 111-12
Sticher, O., 134, 135-6
Stoller, P., 7
Stross, B., 127, 137
subsistence practices in tropics, 77-95
  foragers, 84-90
  root and tuber cultivation, 79-84
susto (ailment), 123
Sutton, M., 111

synaesthetic sound symbolism, 31-49
syntax, 69

takete/maluma figures, 34-7
taxonomic orthodoxy, 4-6
Taylor, P.M., 7
Tebtebba Foundation, 168-70
technical narratives, 185, 186, 190
TEKPAD, 166-7
terra preta do Índio (black soil), 161-2
theory of mind, 61-2
Thompson, B., 89
Torres Strait Islands, pre-European subsistence in, 88-90
totemism, 69
traditional medicine, 117-45
tree nuts, 84-8
trekkers, 102-11
tropics, subsistence practices in, 77-95
   foragers, 84-90
   root and tuber cultivation, 79-84
Tsing, A.L., 112
tuber cultivation *see* root and tuber cultivation
Tyler, S.A., 194

Uznadze, D., 34

Vernonia amygdalina, 132-3
Villa Rojas, A., 126
vocal mimesis, 46-9
Volpato, G., 128

Williams, G., 125, 151
Wilson, E.O., 185
Wilson, S.M., 82
Witkowski, S.R., 124
Wrangham, R.W., 132
Wray, A., 68
writing ethnobiology, 177-96

yam cultivation, 80, 83-4
yellow walnut (*Beilschmiedia bancroftii*), 87-8
Yoruba medicine, 126

Zapotec, 177
Zardini, E.M., 128
Zimmerman, F., 13
zoopharmacognosy, 132-3

Printed and bound by CPI Group (UK) Ltd, Croydon, CR0 4YY
09/06/2025

14686103-0004